AF789633

Bolivia

— The Height of Mineral Collecting —

Written by
Jaroslav Hyršl and Alfredo Petrov

with additional contributions by
Robert B. Cook, Rock Currier, Anthony R. Kampf, Suzanne Liebetrau
Lauren Megaw, Bill Smith, Carol Smith and Terry Wallace

Edited by
Robert B. Cook, Suzanne Liebetrau, Günther Neumeier and Gloria Staebler

Illustrations by
Michael Bainbridge, Daniel Beams, Dan Behnke, Joe Budd, Michael Canoy
Rock Currier, Frank de Wit, Jordi Fabre, Mark Garcia, Tino Hammid
Jaroslav Hyršl, John A. Jaszczak, David K. Joyce, Anthony R. Kampf
Brian Kosnar, Saul Krotki, Jesse La Plante, Gary Moss, Bendegúz Nagy
Günther Neumeier, Tony Peterson, Alfredo Petrov, Mark Pirlo, Jeff Scovil
Terry Szenics and Harold and Erica Van Pelt

Lithographie, LLC • Denver, Colorado

The Political Divisions of Bolivia

(after United Nations Map No. 3875, Peacekeeping Operations, Cartographic Section)

Table of Contents

Quartz & cassiterite
4.3 cm; Huanuni
Dave Bunk specimen
Jesse La Plante photo

Andorite; 6.3 cm
Oruro
Arkenstone specimen
Joe Budd photo

Valentinite; 5 cm
Tatasí; Canadian
Museum of Nature
M. Bainbridge photo

Cassiterite; 14 cm;
Viloco
G. Witters collection
Jeff Scovil photo

Bolivia: A Land of Extremes

*by **Alfredo Petrov***

Lake Titicaca and the Cordillera Real as seen on the road from La Paz to Copacabana; Daniel Beams photo

Are there any mineral collectors who have not dreamed of owning one of Bolivia's famous phosphate specimens? A phosphophyllite twin, perhaps, or a fine vivianite, ludlamite, or vauxite? Or, if tastes are more inclined to the metallic, how about a magnificent andorite, franckeite, or argyrodite? For these minerals and many others, Bolivia has produced the best specimens known for the species, and Bolivia's many mines continue to provide collectors with interesting material.

These Bolivian treasures have been reaching collectors in North America and Europe for a little more than 100 years beginning in the 1880s or 90s, when foreign mining engineers, mainly British, German, and the occasional American, came in search of tin deposits and to reactivate old silver mines. Several Bolivian minerals came to the attention of collectors during this period. Included among these are pyrargyrite from Colquechaca (page 63), probably the all-time most prolific producer of pyrargyrite, and previously unheard of curiosities such as native copper pseudomorphs after aragonite from Corocoro (page 76), strange tin sulfosalts such as cylindrite from Poopó (page 24), and the world's first well-crystallized augelite from Machacamarca.

It is not that no one collected minerals in Bolivia before the late nineteenth century. There is evidence that the Incas collected a few mineralogical treasures in pre-Colombian times: natural mineral specimens including quartz crystals, doubly terminated danburite crystals, and agatized fossils, have been found in tombs of that period. We do not know whether these pre-Colombian people kept the specimens for their beauty, for supposed mystical properties, or both.

The Spanish conquistadores, who arrived in the sixteenth century to pillage the Andes for their silver and gold, did not care at all about mineral specimens. It is heartbreaking to think of how many world-class silver minerals went into the smelters between the early sixteenth and early nineteenth centuries. The very few specimens saved during this period, as for example a couple of large native silvers from Cerro Rico in the collection of Austrian archduke Johann (1782–1859) now housed in the Natural History Museum in Vienna, were probably exported for the purpose of boasting of the richness of the ore rather than for any imagined mineralogical interest.

Even today, Bolivian society reflects the conquistador heritage: the public's interest in minerals is limited almost exclusively to their economic value, with a general lack of regard for the mineralogical value of their country's great diversity of species. The number of serious Bolivian mineral collectors can be counted on the fingers of one hand; nevertheless, numerous Bolivian mines produce remarkable quantities of specimen-grade material, and some miners have learned to supplement their incomes by saving and selling specimens. This issue of *extraLapis English* takes a look at some of Bolivia's mineral wonders, both old and new.

Geography and Geology

Bolivia is a sparsely populated country in the middle of the South American continent. The country is home to about nine million people (about as many as live in New York City) spread over an area almost as large as Texas and California combined. Or, for our European readers, Bolivia is a bit more than three times the size of Germany or twice the size of France.

Bolivia is the poorest country in South America and is the third poorest in the western hemisphere. It has achieved the dubious distinction of being, according to the *Guinness Book of World Records*, the world's most politically unstable country, having changed governments on average more than once a year during its 180 or so years of independence. And far too many of those governments have been brutal dictatorships.

Bolivia, however, is also a nation of social experiment and reform. It was the first country in Latin America to enjoy a land reform revolution (in 1952), the first to have a female president, the first to try neoliberal economics, and most recently, the first to elect an indigenous person to the presidency.

The stereotypical image of Bolivia is a land with great mountain ranges and high cold plains, resembling Tibet. It's a nightmare for those who suffer from altitude sickness. And Bolivia does indeed hold its share of altitude records: At an altitude of 3,640 meters (11,942 feet), La Paz is the world's highest capital city, besting Lhasa, Tibet, by more then 300 meters (1,000 feet). Potosí is the world's highest large city and La Paz's El Alto is the highest international airport; both are at altitudes of roughly 4,090 meters (13,500 feet). Lake Titicaca, South America's largest lake (8,400 km^2), is the world's highest navigable lake (in the 4,000 meter range). Bolivia is home to the world's highest 5-star hotel (in La Paz), the highest rope tow (Chacaltaya, about 5,100 meters), the highest metallic ore mine (Chorolque, up to 5,450 meters), and so on.

But high altitudes only define the southwestern third of Bolivia. Two thirds of the country are tropical lowlands, ranging from hilly to flat. They are covered with alligator-infested swamps, grassy prairies, thorn forests, and Amazonian jungle.

Geologically this vast lowland area (roughly the size of Texas) is an extension of the Brazilian Shield, formed mainly of Precambrian metamorphic and

***Above**: The foothills of Cerro Huañaquino consist of multicolored Cretaceous marls. Near the village of El Molino in Potosí department, one of these strata hosts lapidary-grade stromatolites. Frank de Wit photo, 2009*

***Left**: Swamp Yapacani, Santa Cruz department; Daniel Beams photo, 2007*

plutonic rocks. Most of the Bolivian portion of the Brazilian Shield is buried by up to three kilometers of sediment, which has been washing down from the Andes ever since they rose out of the ocean in late Cretaceous and early Tertiary times (about 60 million years ago).

Fortunately for miners and mineral collectors, there are still islands of Precambrian schists and granites exposed in the hilly areas of central and eastern Santa Cruz department, smaller areas of Beni and Pando departments, and in the rapids of some large tributaries of the Amazon. Although these rocks have been explored far less than their counterparts across the border in Brazil, interesting mineral deposits have been discovered in them over the past 70 years. These include numerous pegmatite districts with beryl and tantalum minerals; several "amethyst" deposits, including the unique Anahí "ametrine" Mine (page 85); Cerro Manomo, the world's largest carbonatite outcrop; and a number of important gold deposits such as Don Mario.

The lowland two thirds of the country are the fastest growing regions, economically. This area boasts a rapidly growing network of roads focused on new agricultural development as the jungle frontier is rapidly pushed back, the harvesting of important timber resources, and the development of energy

Overview of the principal geologic units of Bolivia. Map by Günther Neumeier

resources, all of which will undoubtedly result in the discovery of new mineral localities.

The eastern foothills of the Andes, along the margins of the lowlands, are known to Bolivian geologists as the "Faja Subandina" (Subandean Belt). This belt is composed of numerous parallel ranges of very steep hills, mostly uninhabited and covered with roadless tropical jungles in the north or thorn forests in the south. Much of this region awaits exploration. Biologists, fossil hunters and hydrocarbon prospectors have lots of fun here, but interesting mineral deposits are unlikely. One important exception is the outcrops of metamorphosed evaporites in the Alto Chapare rainforest, well known for their perfect crystals of danburite and boracite and for asbestos mines as well.

The cold highlands of Bolivia are much better known among mineral collectors because of their abundant variety of metallic minerals, especially those of silver, tin, tungsten, bismuth, and antimony, and for beautiful phosphates such as phosphophyllite, vivianite, ludlamite, and vauxite.

The Andean Cordillera runs for 8,000 km (5,000 miles) from Venezuela to Tierra del Fuego along the entire western edge of South America, but it reaches its widest development in Bolivia, where it splits into two parallel mountain chains: the Cordillera Occidental (West Andes), which forms the border between Bolivia and Chile, and the Cordillera Oriental (East Andes). These cordillera are separated by the Altiplano, the high plains (3,600 to 4,500 meters above sea level), which are covered by llama-grazing lands, lakes, and salt flats, including the Salar de Uyuni, which, at 12,000 square kilometers (3 million acres), is the largest salt flat in the world. The famous Bolivian tin belt runs down the western flanks of the East Andes and through island mountains in the Altiplano, from southern Peru to northern Argentina.

The East Andes are composed mainly of a monotonous series of Paleozoic and Cretaceous sediments. At about the latitude of Cochabamba, there is a bend in the Andes; north of the bend the Andes run northwest to southeast, but south of the bend they run north to south. This bend is called *el codo de los Andes,* "the elbow of the Andes". There are interesting differences between the mineral deposits to the north and the south of this elbow.

To the north in La Paz department, large granitic batholiths intrude the Paleozoic sediments. The many tin, tungsten and bismuth deposits here are mainly associated with the batholiths, hornfels, and slate aureoles that are near them. There are no significant silver deposits. The best known minerals from this area are cassiterite from the Viloco Mine (page 72), ferberite and Japan-law quartz twins from Chicote Grande, and more recently, nuggets of native bismuth from the Vilaque river gravels.

South of the elbow, there are fewer batholiths, and most of the minerals are found in Tertiary subvolcanic deposits (i.e., formed under volcanoes, the tops of which are now eroded away). In addition to tin, tungsten, and bismuth, the subvolcanic deposits contain large quantities of silver minerals. The most famous of these deposits are Potosí (page 36), Oruro (page 26), and Colquechaca (page 63). While these are the best known among collectors, there are some 40 similar subvolcanic localities for sulfosalts and phosphates. Some tin-silver and bismuth-tungsten deposits such as Colquiri, Huanuni (page 18), Poopó, and Tasna (page 44) are remarkably similar from a mineralogical standpoint to those of the subvolcanic ores, in spite of the fact that they are not associated with volcanic rocks.

A diverse suite of antimony sulfosalt species, and to a lesser extent bismuth and germanium sulfosalts, characterize the ores south of the elbow, where arsenic sulfosalts are almost nonexistent. North of the elbow, the deposits contain more enargite and other arsenic sulfosalts, somewhat resembling the polymetallic orebodies of Peru.

Apart from the complex tin-silver-bismuth-tungsten ores of the tin belt, the East Andes contain hundreds of gold- and stibnite-bearing quartz gangue veins, and there are literally thousands of galena-sphalerite-chalcopyrite veins associated with siderite and barite gangue. The mineralogy in these mesothermal siderite-barite veins tends to be simple, and they hold little that is of interest to specimen collectors. Notable exceptions are the two unique selenide veins at Pacajake and El Dragon.

In the Altiplano, there is a belt of copper deposits west of, and running parallel to, the tin belt. Secondary minerals are not well developed in these copper mines; the only specimens of collector interest there are native copper pseudomorphs after aragonite from Corocoro.

Unlike Bisbee or Tsumeb, colorful secondary oxidation-zone minerals are generally poorly developed in Bolivia, with very few such species getting good enough to warrant a trip to the specimen market. Since colorful phosphates are characteristic of many oxidation zones world wide, some dealers and collectors have wrongly assumed that Bolivia's vivid phosphophyllite, ludlamite, vivianite, vauxite and other phosphates are oxide zone minerals. In reality, they are always associated with fresh unweathered sulfides in the deepest levels of tin-silver mines. They were hydrothermally deposited and are far from any oxidation zone. Even the exceptional valentinite crystals found at several Bolivian mines and the mimetite and austinite from Lomitas are hydrothermal rather than secondary oxides!

The host rock for Bolivia's hydrothermal ores is usually volcanic, sandstone, shale, or hornfels, which explains Bolivia's dearth of secondary minerals. Carbonate-hosted ores (as in Tsumeb, Laurion, Mississippi Valley Type, etc.) are insignificant in Bolivia; furthermore, calcite, which is abundant in ore veins elsewhere, is remarkably rare in the gangue of Bolivian veins. Without the neutralizing effect of calcite and dolomite, the weathering of sulfides produces extremely acidic solutions, which carry metal ions out of the oxidation zone and inhibit the formation of secondary minerals.

The West Andes are much younger than the East Andes, having formed in late Tertiary and Pleistocene volcanic rocks rather than sediments. The Altiplano formed after the West Andes, which cut off sediment transport westward to the Pacific and leading to the deep accumulation of sediments in the closed basin between the two Andean chains. The West Andes have numerous large deposits of native sulfur with other fumarolic minerals. There are also several mining districts with copper-silver-gold ores in veins and disseminated in altered volcanic rocks. Finally, the salt pans in the Altiplano at the foot of the West Andes hold the world's largest supply of evaporite minerals including halite, ulexite, and lithium-rich brines.

***Top**: The "Faja Subandina" in the eastern foothills of the Andes at Bermejo, near Samaipata in Santa Cruz; Daniel Beams photo, 2005*

***Middle**: Desert of Uyuni, Potosi department Bendegúz Nagy photo, 2006*

***Bottom**: Hot springs in the desert of Uyuni in Potosí Bendegúz Nagy photo, 2006*

Mining, Politics, and Tumult:

by ***Suzanne Liebetrau,*** *a retired teacher, technical editor, and mineral collector currently living in central Oregon.*

Mining and minerals have played an important role throughout Bolivia's turbulent history apparently as far back as ancient times. There is significant archeological evidence that pre-Incan populations smelted silver in the southern areas during 400 to 1000 CE. According to a study published in *Science*, however, silver is rarely found in Latin American archeological sites. The oldest pre-Incan metal artifacts found (in Peru) date from about 1400–1100 BCE (Mayell, 2003).

The Incas ruled the area that now includes Bolivia until Spaniards led by Hernando Pizarrro invaded in the sixteenth century. With their minds and hearts set on the accumulation of precious metals, the Spanish conquerors were pleased to find major Incan silver mining and immediately set about extracting silver for themselves (by means of Incan slave labor) from the rich ground. The city of Potosí, which became the largest urban center in South America, was founded in the 1500s at the site of great silver deposits. Other settlements followed. La Paz (1548), Cochabamba (1574), and Otavo (1604) were all part of "Upper Peru," as the land was known of at the time. Wilson and Petrov have written a good history of silver mining in the Cerro Rico (the Potosí area), published in *The Mineralogical Record* (1999). They note that the Spanish quickly depleted the richest silver deposits. By the mid-1700s, reports to the king indicated that mine workings extended several hundred meters underground and that the deepest mines were experiencing high water flow and poor ventilation.

The eighteenth century saw the rise of the Jesuits, who settled among the indigenous peoples of Chiquitos, in Upper Peru. The Jesuits were charged with improving the spiritual welfare of the locals but insisted that social welfare was equally important.

The Santiago de Chiquitos church in the old Jesuit mission town of Santiago de Chiquitos in eastern Bolivia; Daniel Beams photo, 2006

The photographer and wife, Vanessa, are Christian missionaries working in Bolivia. Daniel travels the country with friends and family taking photographs along the way. He has contributed a number of images to this issue.

King Charles III of Spain by A. R. Mengs, 1761

As a result, their several settlements, called *reducciones*, were profitable: by taking advantage of the fertile land, they taught the Indians to farm and ranch, exporting their goods throughout Upper Peru and beyond. As they became semi-independent, the reducciones attracted both jealousy and political interest. Matters came to a head in 1767, when King Carlos (Charles) III of Spain ordered the expulsion of all the Jesuit missionaries in his realm. Many of the reducciones were ultimately abandoned and were forgotten for over 100 years. Today, several have been resurrected as UNESCO Heritage Sites. (Groesbeck, 2008).

The area now known as Bolivia remained under the control of various Spanish vice-royalties until independence was declared in 1809. During the ensuing 16-year-long War of Independence, mining generally slowed (Moore, 2008). The country's first elected president was General Antonio José de Sucre, who was instrumental in defeating the Spaniards, but Simón Bolívar was considered the true liberator of the new country, which was named for him. Bolívar prepared a constitution, which was adopted in 1826, but stability did not follow the revolution. For the next 100 years, the country was ruled by a series of dictators, mostly military.

Simón Bolívar by Juan Lovera, 1827

Bolivian boundaries and constitutions fluctuated with its leaders. Governmental instability was accompanied by frequent conflicts with neighboring countries. Bolivian territory originally extended to the Pacific Ocean and, thus, was highly desirable. Of course, the areas with rich natural resources were also in demand. Unfortunately for Bolivia, it was usually on the losing side of conflict. Chile won the corridor to the Pacific in 1883, and in 1904, much eastern lowland territory was lost to Brazil. With Bolivia's loss of the coast to Chile came the loss of

a Brief Look at Bolivian History

valuable mineral deposits (copper, nitrates, iodine, etc.); thus, some old labels attribute specimens to localities in "Bolivia" that are currently in Chile. Boundary disputes with Argentina plagued the country until 1925. Paraguay gained control over most of the Chaco Boreal area in the 1930s. Today, Bolivian territory includes less than half the acreage that was claimed in the 1825 revolution. It is the only landlocked Andean country and is one of the most unstable of Latin American countries.

Throughout these turbulent times, mining continued. The collapse of the international silver market at the end of the nineteenth century coincided with a growing demand for tin, which, until that time, had been considered an unimportant byproduct of the silver mines. The subsequent tin boom gave rise to a new set of entrepreneurs. The tin mining industry was soon dominated by three major corporations, led by three powerful men: Simon I. Patiño, Mauricio Hochschild, and Carlos Aramayos. These "Tin Barons," along with influential landowners and military leaders, controlled Bolivian politics until about 1952, although by about 1913, the global demand for tin slumped (with temporary rises during World Wars I and II).

In 1932, the unfortunate "Chaco War," a conflict on the border with Paraguay over supposed deposits of oil (which may never have actually existed) resulted in the loss of over 60,000 soldiers—most of whom were indigenous conscripts—out of a population of 2 million. In the aftermath of the conflict, mistrust and anger were directed toward the ruling class and their political system. New, more radical ideas and parties began to rise in influence.

The most important of the new parties was the Movimiento Nacionista Revolucionario (MNR), which advocated nationalization of the mines. In the 1940s, the MNR was supported by the powerful Federación Sindical de Trabajadores Mineros de Bolivia (Federated Union of Mineworkers of Bolivia, or FSTMB), which added significantly to the strength of the MNR.

In 1952, the MNR took over the Bolivian government. Under the MNR, the state nationalized the holdings of the three Tin Barons, and two-thirds of the tin mining industry came under government control. The mines were managed by a new, state-run mining corporation, the Corporación Minera de Bolivia (COMIBOL). This reform movement lasted until 1964, when the MNR was defeated and deposed by military force. Between 1964 and 1989, instability reigned once more: of 19 presidents, only 2 completed full terms of office!

The Aramayo Bismuth Mine and the famous Chorolque Peak (6,000 meters) from a circa 1910 postcard. The photo was taken by Luis D. Gismondi, La Paz.

Günter Grundman collection

Amid this turmoil, a growing movement among farmers, displaced miners, priests, and union leaders resulted in organized committees marching for individual rights. In 2005, indigenous Bolivians turned the tables on their restless and violent history of subjugation by electing President Evo Morales, former leader of the country's coca growers. Morales, of Indian descent, promised a voice to that major component of Bolivia's population. Symbolic of this mood, 36 native tongues (with Spanish) spoken in Bolivia were deemed "official" in the constitution drafted by the new government. Morales's constitution became official in January of 2009.

Morales's term has not been a calm one, however. Differences among the indigenous people and those of European descent still present obstacles to progress. Moreover, today, as in historical times, Bolivia's natural riches still drive political and economic concerns. The exploitation of tin has made Bolivia one of the world's major suppliers. In addition to the tin, silver, and rich resources of oil, salt, and zinc, Bolivia possesses South America's second largest proven natural gas reserves. Morales has nationalized both the natural gas and the oil industries and has given the indigenous peoples control over natural resources in their territories. However, his attempt to nationalize one of the largest tin mines, Huanuni (page 18), resulted in bloody battles between factions of miners: those from the cooperatives and those earning salaries from the state. As a result, cooperative mining was declared illegal. The cooperative miners in Huanuni were offered jobs working for the state, and about three-

Butch Cassidy and the Sundance Kid

*by **Lauren Megaw***

The reality behind Butch Cassidy and the Sundance Kid is shrouded in legend. The two were bandits in the United States, and when they moved to the southern hemisphere, they became famous outlaws who wreaked havoc and eventually brought about their own demise. Distinguishing fable from fact is a test of how much faith one places in the anecdotes of their adventures that have been passed down, filtered through the 1969 blockbuster Hollywood portrayals by Paul Newman (as Butch Cassidy) and Robert Redford (as the Sundance Kid). Although the movie is a fairly good rendition of their story, quite a few changes were made for the sake of moviegoers, glossing over the fact that they were ruthless, coldblooded killers with the blood of dozens on their hands.

The Wild Bunch
John Swartz photo, 1900

The man known as Butch Cassidy was born Robert Leroy Parker (April 6, 1866, Beaver, Utah). The "Sundance Kid" was christened Harry Longabaugh (1867, Mont Clare, Pennsylvania). Both became part of the infamous "Wild Bunch," a gang that perpetrated a string of train robberies throughout the western United States. The Wild Bunch knocked over trains for the payroll they carried, and then escaped to their "Hole in the Wall" hideout. With the law closing in on them, Butch and Sundance dashed to New York and boarded a boat to Argentina with Sundance's significant other, Ethel Place (also known as Etta) in tow.

The three became homesteaders in Argentina, where they laid low until the Pinkertons, who had been tracking them, began to close in. Before long, Ethel Place was on her way back to the United States; the men having been accused of a robbery they probably did not commit. Butch and Sundance fled first to Chile and then to Bolivia, where they began planning what was to become their finale: a magnificent series of heists focused on the rich silver and tin mines of Bolivia.

The two men decided to capture the payroll of the Aramayo Mining Co., operators of the rich San Vicente silver mine. (Today, the San Vicente Mine is run by Pan American Silver Corporation. Thirty-eight different mineral species are known from the district, and many, such as franckeite, make fine specimens.) They seized the payroll on November 3, 1908, jumped the man with the payroll and ran off with the loot and the mule. The men hid in the town of San Vicente, but the denizens reported the two Americans to the authorities (after one of them had recognized the "Q" brand on the mule, which they had been alerted to look for by the authorities). On the night of November 6, 1908, their hideout was ambushed by the Bolivian Army--- they were besieged. Butch and Sundance attempted to defend themselves, but well into their O.K. Corralesque clash, witnesses attested to screams coming from their hideaway; a shot from inside was followed by silence; then came a second shot.

The army made its way into their hideout, and as the story goes, they found the bodies of Robert Leroy Parker and Harry Longabaugh. Longabaugh appeared to have been shot and Parker seemed to have committed suicide. A story circulates that Butch committed suicide because he was out of ammunition, but other stories have several hundred rounds of ammunition remaining with the dead outlaws.

The two men were supposedly buried in an unmarked grave, but neither it nor their bodies have ever been found. A grave reputed to have been theirs in the nearby town of Tupiza (which makes significant income from Butch and Sundance tourists) was excavated in the early 1990s, but DNA tests proved that it did not contain the bodies of Parker and Longabaugh. One school holds that Butch and Sundance survived Bolivia and returned to the United States, but only circumstantial evidence supports those claims.

We are unlikely to ever know what really happened to Butch Cassidy and the Sundance Kid, but perhaps it is fitting that the story of two men who were both detested and revered during their lives, and who have been elevated into popular icons by dime novels and Hollywood, remains a tale shrouded in mystery.

Miners demonstrate in the streets of Oruro
Frank de Wit photo, 2009

quarters accepted (McCarthy, 2009).

Some argue that state ownership is the only way to ensure that mining is carried on safely and responsibly and that the mine's life is maximized. Others feel that the new salaries will result in reduced profits. The Bolivian government is affecting the status of Bolivia's other mines as well, levying substantially increased taxes on foreign-owned investments (McCarthy, 2009). Foreigners interested in extracting Bolivia's natural riches without a profit-sharing agreement are strongly discouraged (Jackson, 2009). As James Read notes in his excellent summary of Bolivian history, the election of an indigenous president may herald a "new dawn" for Bolivia, but the country's turbulence persists. "Poverty, underdevelopment, and ungovernability" (Read, 2008) continue to plague Bolivia; it remains a battleground for natural riches and political domination.

Road Trip Bolivia

by ***Rock Currier***

La Paz sits in a bowl; El Alto, La Paz's airport, is up on the rim. To become acclimated to the altitude, travelers descend into the bowl of the city for a few days after flying into Bolivia.

Bendegúz Nagy photo, 2006

Bolivia is a tough country; one that does not easily give up its fine mineral specimens. Most of its mines are located in remote, treeless, mountainous areas at altitudes in excess of 4,000 meters. The roads—if that's what you want to call them—are bad by American and European standards, really bad. Still, after you have been there a few times, the place grows on you, and you want to go back and try your luck again, believing that in that next incredibly remote place, treasure most certainly awaits.

If you travel to Bolivia to buy specimens, you have to work harder for them than you would in other countries because the specimens are not concentrated in markets or with regional dealers; rather, they are scattered among individual miners, who can be found at the mines, or among a few small dealers. In other countries, mineral specimen markets have developed in specific towns and cities; for example, there are mineral markets in Soledade and Governador Valadares, Brazil, or in Changsha and Guilin, China. In these places, you can run from dealer to dealer and do most of your buying. Of course, if you don't want to travel to far, inhospitable places, you can shop for mineral specimens at the granddaddy concentrator of them all: the giant annual gem and mineral show in Tucson, Arizona.

In Bolivia, buyers must travel all over the place to find specimens, enduring all sorts of difficulties along the way. Sometimes, for example, they get close to their destination only to find their way blocked by a river or that there is no way to get gasoline. Perhaps a buyer heads to a place where the miners happen to be pissed off at the government, the mining companies, or their lot in life and have rolled a huge boulder onto the road to prevent anyone from passing. Multiple flat tires and mechanical breakdowns could impede a buyer's progress, or a snowstorm might block the roads. Any number of things can, and will, crop up to let you know that you are in a place where anything can happen.

Several trips to Bolivia remain vivid in my memory. My usual routine was to fly into La Paz and rent a vehicle. More accurately, I would fly into El Alto (altitude 4,169 meters or about 13,600 feet), feeling a little wobbly as I got off the airplane and wondering whether I was going to need one of the little oxygen cylinders that the airline people keep handy. El Alto is located on the Altiplano, a plateau that runs roughly north and south for several hundred miles through Bolivia, Chile, and Peru. After

Right: Sometimes you just can't get there without dynamite! A rock slide blocks a Bolivian road preventing Michael Canoy and Alfredo Petrov from passing (How did they turn around?).
Michael Canoy photo

Author Rock Currier took this photo as he waited to cross the water-covered road that nearly swept away him and his rented VW Bug.
Rock Currier photo

arriving, I typically got onto the bus that travels over rim of the canyon and down a couple of thousand feet into the valley where La Paz is located. As in other high-altitude places, the expensive real estate is not located high up where the view is best; it is located lower down where there is more oxygen, warmer temperatures, easier access to water, and even a few real trees. I always found enough to do for a couple of days in La Paz to keep me busy and allow my body to adjust to the altitude before renting a vehicle to head up to the mines. In a pinch, you could get by with a Volkswagen Bug, but you really need a 4x4 to negotiate the roads to the mines.

On one particular trip, I was short of funds; so I rented a Bug and, in retrospect, realize the trip was a sort of suicide run. I made the mistake of not checking to see whether the car was in good condition. (Don't do this!) But soon after renting the Bug, I was winding my way up the canyon wall up to El Alto, past the airport and enjoying a nice smooth ride south across the Altiplano south to the old mining center of Oruro. Everything was great. Perhaps the stories I had heard about the terrible roads had been exaggerated.

Oruro (page 26) was a rough little town that had grown up around the great Itos and San José mines, which had operated since Spanish times. The pickings were slim, and I had trouble locating miners who had specimens to sell. The Universidad Tecnológico de Oruro had a collection that looked like it had seen better days, although they did have one good valentinite specimen, which, on a later trip, I found had vanished. There was nothing left to do but to head out to the mines and see if I could scare up some specimens to keep my fledgling mineral business alive. It was either that or find an honest job. Right on the southern edge of Oruro, the pavement was replaced by a morass of deep mud and ruts that even big trucks were having a tough time negotiating. I managed to keep my wheels on the high centers and without too many scraping sounds and thumps on the undercarriage, got past the worst spot. From there to Huanuni, the road was not too bad. At one point, there was a dip in the road that had a little river running across it, but my trusty little Bug waded through it without much trouble.

At Huanuni, I started knocking on miners' doors, but again the pickings were slim. To hit mining camps cold is usually not a terribly efficient way to get specimens, but sometimes it is the only way to get started. Put yourself in the position of the miners: Here comes some foreigner knocking on your door wanting to know whether you will sell him some rocks. To many miners, this is just strange. Perhaps the visitor is a crazy person. Perhaps he works for the mining company and is trying to find out whether you are stealing ore from the mine. In the miner's position, it's just better to say "No."

On the way from Oruro to Llallagua (page 50), I noticed that my brakes were not working all that well. Whenever I stepped on them, there was an ominous sound of metal on metal and the Bug would pull sharply to the left. In Llallagua, my efforts were nearly a complete bust. The miners lived all over town—it was really a city more than a town—and I could locate very few specimens to buy. I found I needed to find relatively flat places to park because the parking brake wouldn't hold the Bug on a slope.

I decided to cut my losses and try to make it back to Oruro for the night. All went well until I got back to the little river crossing a few kilometers from Oruro. The sky was clear, and it was near sunset. There were a few cars and trucks stopped on both sides of the river. I thought that was a little strange, and although the water looked a little higher than it had been when I had crossed in the morning, it didn't look all that much higher; so I just drove across. The first thing that got my attention was the bow wave of water that washed up on my windshield. Then I noticed that the Bug seemed to be floating a little and scraping a bit sideways across where I thought the road was. There wasn't much I could do but gun the engine.

Periodically the heavier rear of the Bug would sink down and hit the road, kicking the car forward a bit until the rear would sink back down and kick me further across. My knuckles were white from their death grip on the steering wheel, but amazingly enough I made it across and up the other side. A few of the people who were stranded on the riverbank clapped and cheered. I pulled over and stopped till my breathing and heart rate slowed and I was able to loosen my fingers from the steering wheel. That night in Oruro, I gave some thought to what it must have been like for Sam Gordon of the Philadelphia Academy of Science when he was travelling to Llallagua on his motorcycle in the early part of the twentieth century and pondered just what it was that I was trying to accomplish in the mineral specimen business.

The next day I headed back down that nice smooth road to La Paz. Nothing bad could happen now, I thought. Then about 12 kilometers out, the accelerator went flat to the floor, and with the engine idling, the Bug slowed to a stop. The accelerator linkage to the engine that ran under the car had snapped. Some of that scraping and thumping on the undercarriage must have done it in. The engine was working just fine, so I took a rock from the side of the road and jammed it in the carburetor linkage until the engine was racing pretty good. Then I got back in and using the clutch and gears, was soon humming along down the road to La Paz.

When I stopped to congratulate myself on how clever I was, I realized that I would soon face traffic and the long steep road from El Alto down into La Paz with questionable brakes. Before plunging into the city, I replaced the rock in the carburetor linkage with a smaller one so the engine would run slower. Then, working the clutch and the gears like a maniac, down I went. Periodically I had to stop for traffic, and that was a real problem because the brakes didn't work very well. Even when fully engaged with metal grinding and the parking brake yanked up, the poor little Bug would barely stop. With more white knuckles on the steering wheel and emergency brake lever (and an elevated heart rate), I made it. That evening, I thought about what I was going to tell my customers when they complained that my prices were too high. Better to move on to Brazil or Peru where the pickings were easier.

Not all my trips to Bolivia were as tough, and there were a couple that were almost magical. I remember a drive one afternoon out to the tin mine at Viloco (page 72), famous for what are probably the best cassiterite specimens in the world. The road curved along a canyon wall where the occasional ore car rails came directly out of an adit and across the road to a cliff face hundreds of meters high. The road continued and curved across the end of a high valley just below the toe of a glacier where women were washing clothes in the melt water. The vistas were spectacular; probably the finest of any trip I have taken in the Andes.

On another trip, I visited the tungsten mine at Tasna (page 44) and bought spectacular twinned ferberite crystals right out of ore sacks, paying almost nothing for them. On yet another trip, while hiking on the isolated island of Inca Huasi in the middle of the largest dry salt lake in the world, a wild eagle befriended me and landed on my shoulder. Who would not want to go back to a place like that?

Known as the most dangerous road in world, the road from La Paz to Coroico, in the Yungas region, offers glorious vistas but sees horrific accidents—lots of them. The road takes travelers through several climates from La Cumbre, the pass at the top of the Andes (ice and snow) to Coroico (bananas and monkeys).

Daniel Beams photo, 2008

Buying and Collecting: Local Customs and Practicalities

*by **Alfredo Petrov***

A woman sells her wares in one of Bolivia's many local markets, where there is always a chance of finding the stray mineral specimen.

Bendegúz Nagy photo, 2005

Although Bolivia is a mineral-rich country, the prospects for field-collecting top-quality specimens during a brief visit are low. Purchasing specimens from local dealers and miners is an option but can be a time-consuming process because, as Rock Currier noted in the previous article, there is no central mineral market in Bolivia and the few local dealers are scattered across different cities. Furthermore, the prices that local dealers charge are not necessarily less than those charged in Tucson or Munich, unless one is buying wholesale quantities. One can find mineral specimens in a few shops, especially in La Paz, but here again prices are generally high.

El Tio, "the Uncle," is festooned with gifts, which are offered by miners in exchange for good luck in the mine. Some version of el Tio is found in every underground mine in Bolivia.

Alfredo Petrov photo

Nice, reasonably priced specimens can occasionally be found in local handicraft or tourist markets, albeit from sellers who haven't the foggiest notion about correct identifications or localities. When visiting mines, one can sometimes buy specimens directly from the miners at super bargain prices. While miners generally have no idea what collectors are looking for, a few days of rummaging through boxes of dusty massive ore samples and broken crystals can net a world-class specimen.

The old-fashioned idea that everyone is responsible for their own safety prevails in Bolivia! Consequently, collectors are reasonably free to climb around old dumps and even to explore abandoned underground workings. On the dumps, break open lots of boulders and look for small vugs. The matrix in many of Bolivia's most famous localities, including Potosí, Oruro, and Llallagua, tends to be hard massive pyrite or extremely hard silicified volcanic rock; so bring a heavy hammer. Most vugs contain only quartz and pyrite, but with patience, it is still possible to find more interesting well-crystallized species, mainly micros. Typically, andorite, franckeite, zinkenite, and augelite are found in Oruro; and wurtzite, tetrahedrite, jarosite, and others in Potosí; cassiterite, monazite and other phosphates in Llallagua; and cassiterite and scorodite are available in the dumps at Huanuni.

Visiting active underground workings with miners has even become popular among tourists in Potosí

Left: *This 23.5 cm tall fancy, double hook, iron lamp from Cerro Rico (date unknown) includes a cross, mining tools, and coca leaves (for fertility).*

Right: *This ornate, 24.1 cm tall silver-alloy spout lamp from Cerro Rico (date unknown) depicts a crucifix, which signifies Christianity, a bull for strength, and a condor for fertility.*

Both lamps are in the collection of Tom Hughes. Jeff Scovil took both photos.

Information on the lamps was taken, with permission, from Pohs (1995).

(see page 36). A typical underground tour takes three hours and costs 5 to 20 dollars. It is also possible to meet cooperativa miners early in the morning as they enter their adits and ask to accompany them to their working face. In such cases, it is customary to bring small gifts, such as cigarettes, brandy, coca leaves, or dynamite.

Underground, the only mineral specimens that are generally visible are post-mining water-soluble iron and aluminum sulfates. Both Potosí and Oruro produce enormous quantities of melanterite, copiapite, halotrichite, alunogen, and many others. Underground tourists in Potosí are often impressed by the abundant sea-green translucent stalactites of melanterite (usually misidentified in Bolivia as "chalcanthite"), but these dehydrate to a white powder within a few hours after removal from the mine. Cerro Rico has produced well-crystallized blue chalcanthite, orange sideronatrite, and olive-green voltaite; while Oruro yields well-crystallized potassium alum, green melanterite, and violet coquimbite. These unstable species are best preserved in black plastic film containers that are kept in the refrigerator, where the temperature, light conditions, and lack of air circulation resemble their natural environment.

Before venturing underground, it is customary to leave a cigarette, small cup of alcohol, or a few coca leaves in front of the statue of *el Tio*, "the Uncle," the horned devil found in every mine in Bolivia. El Tio owns the minerals and must be compensated for their removal; otherwise, one risks one's ore vein pinching out or death by collapsing rocks in the stopes. Even the smallest underground workings have their Tios, which range from small, simple, mud figures to life-size psychedelically decorated statues. The miners, most of them nominally Roman Catholic, believe that the Christian god rules above the surface, while El Tio rules underground.

In the early days of Bolivian mining, pan lamps were placed above the statue and offerings were hung on the lamps or placed in the cups. The lamps pictured here were used in Cerro Rico. Candles or lamps of a simpler style later replaced these fancy lamps (Pohs, 1995). Such lamps are no longer found in Bolivian mines, but the tradition of honoring el Tio lives on.

By the way, miners are not superstitious about having their el Tio photographed. The author has only once been denied permission to snap a picture of a Tio, and that was not because of superstitions but because the miners were embarrassed that their Tio had lost its horns. Apparently, a few days earlier the Tio had had horns formed of dried mud, which dissolved when revelers poured beer over its head during his inauguration ceremony.

Top left: *Ludlamite from Huanuni; 5 cm tall*
Fine Minerals International specimen; Jeff Scovil photo

Above: *Andorite from the Itos Mine, city of Oruro; 4.1 cm wide*
Terry Wallace collection; Jesse La Plante photo

Below: *Vivianite from Morococala; 4.7 cm long*
Francis Benjamin collection; Jeff Scovil photo

Lower right: *Cylindrite from the Santa Cruz Mine, Poopó; 3.6 cm tall*
Paul Stahl collection; Jeff Scovil photo

Around Oruro

Below left: *Lake Poopó was photographed in 1991 by the crew of the space shuttle. When full, this salt lake covers 2,500 square kilometers. It is located more than 3,650 meters above sea level and has an average depth of only 3 meters. NASA photo*

Below right: *Signs across the Altiplano warn of llamas crossing. Frank de Wit photo, 2009*

Bottom: *Along the Altiplano*
Daniel Beams photo, 2007

Oruro Department

Map by ***Günther Neumeier***
Text by ***Alfredo Petrov***

Oruro is the only department (state) in Bolivia that does not possess lowlands; the lowest spot in the department is the northern shore of the Uyuni salt flats, at 3,600 meters above sea level. The famous San José and Itos tin-silver mines, and several lesser known mines, are located within the city limits of Oruro, the like-named capital of the department. The homonymous department and city names lead to plenty of label confusion: Could this specimen be from the city? Or is it from one of the other important mines in the state, such as the Huanuni, Morococala, or one of the several mines in the Poopó district?

The people of Oruro endure the frigid climate, in mostly unheated buildings, but they are known in Bolivia for their warm hearts; they are among the friendliest of the country's inhabitants.

Vivianite and ludlamite from Huanuni Dalence province Oruro department; 10.7 cm tall; Cal Graeber specimen; Jeff Scovil photo

Huanuni: Bolivia's Tin Town

by ***Jaroslav Hyršl*** *and* ***Alfredo Petrov***

History

In the early nineteenth century following the chaos of the wars for independence, Bolivia's silver mines suffered great decline and many mines were paralyzed. Around the middle of the nineteenth century, mine owners began to take an interest in tin, and by the 1890s and early years of the twentieth century, many old silver mines had reopened to produce this "new" white metal. The first mine in the town of Huanuni in Dalence province, Oruro department, opened in 1880, making Huanuni home to the oldest of Bolivia's big tin producers. Huanuni temporarily became a silver town again from 1912, when a rich silver vein (0.5 percent silver) was discovered in the old Porvenir Mine (now completely buried by dumps from younger Huanuni tin workings). Silver production lasted until 1928. Interestingly, discrete silver species did not occur in the deposit; the silver was contained in the antimony sulfosalts jamesonite, boulangerite, franckeite, and cylindrite.

The mines in the Huanuni district have, through the years, been run as separate mines, then combined, redivided, and combined again. They were first exploited by Englishmen, then later acquired by the tin baron Simon Patiño before they were passed to the government-run Corporación Minera de Bolivia, (known by the acronym COMIBOL) with the 1952 revolution. Huanuni may have been the richest tin deposit of all time, exploiting, in some places, veins up to 6 meters thick of almost pure cassiterite! Production of high-grade ore continued at least into the mid-1979s when veins of somewhat vuggy cassiterite up to 1.2 meters wide were still being mined.

COMIBOL, like state-run enterprises everywhere, was hampered by bureaucracy and corrupt politicians. The state-owned mines, thus, lost money and needed government subsidies to survive. But Huanuni was so rich that even COMIBOL was capable of making a profit there! As COMIBOL declined, most of the mines were transferred to miner-owned cooperatives and left to sink or swim. Not Huanuni, however. The government kept this mine, because of its profitability, until a few decades ago, when it was finally sold to the British company Allied Deals. Allied Deals abandoned Huanuni in 2000, and it reopened as a COMIBOL mine in 2001.

Cooperatives continued to operate small mines around the edge of the district. Some miners worked fine-grained cassiterite in the upper oxidation zones on Mount Pozokoni; some washed alluvial cassiterite in river sand (much of which leaked out of the inefficient upstream mill), and some searched for silver-tin ores in the outlying sulfosalt zone. In 2006, pitched battles between coopertiva miners (using old rifles and homemade dynamite bombs) up on Cerro Pozoconi and salaried COMIBOL miners at the bottom of the mountain resulted in 16 deaths, hundreds of injuries, and dozens of destroyed homes. Huanuni is now under military control to prevent further violence, and the cooperativistas are under pressure from the government to leave Huanuni and mine distant claims that the government is putting at their disposal.

South America, Huanuni
In its heyday Huanuni was, as the caption on this 1930s era stereo photograph (only one side is shown) reads, "The largest tin mine in Bolivia. 3,800 meters altitude."

Neue Photograpische Gesellschaft (Berlin) photo from the collection of Günter Grundmann

Geology

Although Huanuni's mineralization is remarkably similar to that of Llallagua, the Huanuni veins are emplaced in slightly metamorphosed Silurian and Devonian sandstones and shales. Unlike Llallagua, there is no volcanic stock, although one can assume the presence of an igneous intrusion at depth, and porphyry dikes are known in the area. Tin-bearing veins occur over an area of 1.5 by 1.2 kilometers centered around the peak of Mount Pozoconi, elevation 4,524 meters. Outside the tin-bearing area,

Lower left: *Pyrrhotite on siderite; 6.5 cm wide; Jaroslav Hyršl collection and photo*

Upper left: *Arsenopyrite; 5.5 cm wide; Jaroslav Hyršl collection and photo*

Right: *Vivianite; 6.3 cm tall; Martin Zinn collection; Jeff Scovil photo*

mines working belts of lower temperature deposits have produced sphalerite and argentiferous lead-tin-antimony sulfosalts, and other mines have produced stibnite with hübnerite.

An interesting phenomenon at Huanuni is reverse enrichment: unlike the secondary sulfide enrichment commonly found in silver and copper deposits all over the world, Huanuni's reverse enrichment occurred when the sulfides were completely oxidized and leached away, leaving large caves in Mount Pozoconi. The insoluble cassiterite collapsed to the bottom of the veins and formed the incredibly rich cassiterite breccias for which Huanuni is famous. The huge amounts of sulfuric acid that were liberated by this process attacked the silicates in the surrounding shale and hornfels country rock, resulting in caves with karstlike stalactites and wavy curtains formed of **opal**, **limonite**, and **jarosite**.

Minerals

Sulfides and Sulfosalts

Pyrite is a common mineral, and octahedra modified by pyritohedron faces were mentioned by Cook (1975). "Roses" of many slightly tilted pyrite cubes were found recently, as were lens-shaped pyrite pseudomorphs after siderite. **Pyrrhotite** must have been common at one time, but most of it has been replaced by pyrite and marcasite. A very unusual form of probably the youngest generation of pyrrhotite was found in 1997: radial brown balls on siderite and similar small balls on and in ludlamite. They do not have much of a metallic luster but can be easily recognized with a magnet. Well-formed green-gray twinned **stannite** crystals up to 1 centimeter are rare in Huanuni. **Greenockite** is common as a bright brick-red thin crust on cassiterite, quartz, or siderite. In hand specimens, greenockite is difficult to distinguish from similar common crusts of **hisingerite**. Recently collected **arsenopyrite** is among the best in Bolivia. It forms druses of very lustrous short prismatic crystals that reach widths of 1.5 centimeters. **Franckeite** was common in banded veins in the Porvenir and Maria

Francisca mines. It sometimes forms forms globular aggregates to about 2 centimeters. The mineral was exploited for both its silver and tin content.

Phosphates

Vivianite from Huanuni does not reach the lengths of the longest specimens from Morococala, but these crystals are probably the world's gemmiest, although their perfect cleavage prevents their being cut into gems. It forms thinner prisms than that from Morococala. Some older specimens consist of very thin, even acicular, terminated crystals, reaching 5 centimeters, in fan-shaped aggregates. Small bubbles on the surface of an older vivianite generation cause rare black phantoms inside some crystals.

Ludlamite is the best known mineral from Huanuni, which produces some of the finest examples of the species. Ludlamite was first found in 1993, and subsequent finds have been few. Sharp green crystals can reach 4 centimeters in length. Ludlamite usually occurs in association with long-prismatic vivianite crystals on siderite and pyrite and is sometimes included with millimeter-size pyrrhotite rosettes.

Aheylite, a rare member of the turquoise group, described by Foord and Taggart in 1998, is the first species for which Huanuni is the type locality. At the time that mineral was described, the type specimens were erroneously attributed to the "Miraflores Mine." The type specimens have since been examined and are widely accepted as being from Huanuni. Aheylite is found as pale blue-green to blue-white balls, reaching to 2 millimeters, on quartz with cassiterite, sphalerite balls, wavellite crystals, and variscite.

Wavellite is far less common from Huanuni than from Llallagua, but small white balls are found occasionally. The best specimen consists of an almost 3-centimeter-wide white ball on a sulfide matrix, alongside colorless fluorite cubes.

The late Gene Foord distinguished two types of **variscite**, (variscite-M and variscite-L) and believed them to be two different species. He found both types in Huanuni.

Lower left: *Wavellite on pyrite; 2.7 cm wide; Kosnar collection; Brian Kosnar photo*

Upper left: *Aheylite on cassiterite; field of view 1.5 cm wide; Caltech collection; Mark Garcia photo*

Upper right: *Ludlamite 8.5 cm wide Wilensky specimen Jeff Scovil photo*

***Lower right**: City of Huanuni; Jaroslav Hyršl photo (1995)*

***Upper left:** Elongated crystals definitively identify Huanuni as the locality for this 3.3 cm wide Bolivian cassiterite, collected in the mine in the 1970s. Arkenstone specimen; Joe Budd photo*

***Upper right**: Nikischerite from Huanuni The largest crystal is 0.9 cm wide. Ron Pellar collection; Jeff Scovil photo*

***Lower left**: This 8 cm tall valentinite from Huanuni was probably collected during the Bolivian expeditions of the 1920s and 30s before being acquired (with Vaux funds) by the Philadelphia Academy of Natural Sciences. Chris Stefano collection; Jesse La Plante photo*

Other Minerals

Cassiterite is the main ore mineral at Huanuni. Its habit easily distinguishes it from cassiterite from other Bolivian localities. Huanuni cassiterite forms lustrous black crystals, which can reach 2 centimeters. Crystals typically consist of elongated prisms terminated by a complex of low pyramids. According to Cook (1975), these are cyclic twins of four individuals. More typical coffee-brown crystals lacking significantly developed prism faces occur in the massive cassiterite veins. Better specimens consist of plates of free-standing prismatic cassiterite with white quartz crystals. Huanuni is noted for cassiterite (variety wood tin) and has been the object of research into the details of its formation.

Siderite is widespread. It forms yellow to brown crystals on quartz that reach 1 centimeter. The locality produces a variety of habits of siderite, including lenslike, steep, pointed crystals.

Radial groups of **valentinite** crystals up to 5 millimeters with boulangerite and cervantite were classic specimens from the Porvenir Mine in 1920, but are completely unobtainable now. A cave near the top of Cerro Pozoconi at one time yielded excellent stalactites of white opal. Unfortunately, very few of them were saved.

Cronstedtite is found as black triangular crystals up to about 1 cm long that superficially resemble tourmaline. They sometimes form hemispheres on siderite. There has been a recent abundance of drusy cronstedtite crystals on hisingerite crusts.

Nikischerite ($NaFe^{2+}_6Al_3(SO_4)_2(OH)_{18} \cdot 12H_2O$) is the second species for which Huanuni is the type locality. Nikischerite forms green balls of platy crystals. It is the ferrous analogue of shigaite and was named for Anthony "Tony" Nikischer, dealer of rare mineral species. Nikischerite crystals reach several millimeters, usually aggregated in spheres from 5 to 14 millimeters across. Associated species include vivianite, siderite, pyrite, cronstedtite and amorphous zinc sulfide. It has been found sporadically since the early 1990s but has been overlooked by collectors primarily interested in the associated vivianite! The mineral is unstable and slowly oxidizes to yellow-brown.

Cronstedtite from Huanuni; the field of view is 4.2 cm wide; Rock Currier collection; Jeff Scovil photo

Collecting

Huanuni has been historically far less interesting to collectors than has Llallagua. During the period of COMIBOL control, collecting of specimens by miners was severely discouraged and the purchase of such specimens considered a crime, as one unfortunate dealer found out in 1976, but the locality has produced excellent specimens of ludlamite and vivianite over the last decade. These have generally been immediately purchased as a lot from mine engineers. To find a single nice vivianite or ludlamite in Huanuni by oneself would be tantamount to a miracle, but collectors can climb over the huge dumps and "glory holes" on Mount Pozoconi and find nice specimens of crystallized jarosite, scorodite, and other oxidation zone species, as well as vugs with drusy cassiterite, tourmaline, and such oxidation-resistant minerals.

Nowadays, after the recent political conflicts in Huanuni, visitors are certain to be harassed and possibly searched by nervous soldiers or police, but this situation may calm down with time.

Huanuni Cave Formations

*by **Alfredo Petrov***

Enormous cassiterite-pyrite veins once traversed the shale/hornfels at Huanuni. In time the pyrite decomposed, liberating sulfuric acid (H_2SO_4), the loose cassiterite collapsed to the floor of the veins, thick masses of cassiterite breccia were formed, and large openings (caves) at the tops of the veins were created. The sulfuric acid attacked the surrounding shales, and opal stalactites and stalagmites formed in the caves.

Cavers are generally familiar with the calcium carbonate speleothems produced when carbonic acid acts on limestone. But when sulfuric acid reacts with silicate rocks, it produces a chemically more diverse suite of minerals including goethite, jarosite, diadochite, and the spectacular opal shown here, which probably formed as a stalagmite, whose growth was periodically interrupted by mud deposited on the cave floor. Opal speleothems also form in the Colavi Mine, Potosí department.

Opal stalagmite from Huanuni; 8 cm long; Stuart Strife collection; Jesse La Plante photo

The Poopó District

by ***Jaroslav Hyršl*** *and* ***Alfredo Petrov***

Upper left:: *The Santa Cruz Mine tailings*
Lower left: *A miner in a Santa Cruz Mine stope*
Right: *Monument to the mine at Poopó; the chimney is part of a now-defunct smelter*
Frank de Wit took all three of the photos on this page in 2009. Frank and author Alfredo Petrov lead mining and mineralogy focused trips to Bolivia.

From an economic point of view, the tin-lead-silver-zinc mines of the Poopó district, located on the eastern edge of the Altiplano, 55 kilometers south-southeast of Oruro, would barely rate a mention, but they are famous among collectors as about the only source of the mysterious mineral cylindrite. The locality also produces a few other rare tin sulfosalts and sulfides, available only from a few localities in Bolivia. This district is one of very few in the world in which tin sulfides are significant tin ore minerals, rather than cassiterite, the much more abundant tin oxide. The name Poopó tends to incite giggling among English-speaking youngsters, which diminishes when correctly pronounced *Poh-POH.*

Tin Sulfosalts

Cylindrite, $(Pb,Sn^{2+})_8Sb_4Fe_2Sn^{5+}_4S_{27}$, is popular among collectors because of its long metallic cylinders (hence the name), which resemble pencil leads. No other macroscopic mineral exhibits cylindrical crystals, and the structural origins of this unique shape have been under intense study for decades. The species was originally found in the Santa Cruz Mine (the type locality) in the town of Poopó and was described as a new mineral in 1893. Larger crystals were later found in the Trinacria Mine, 7 kilometers south of Poopó. Crystals from Trinacria consist of cylinders up to 1 centimeter in diameter and 6 centimeters long. Much thinner cylindrite crystals have been found at the San Francisco and Rosario mines, both of which work extensions of the same set of veins as the Santa Cruz Mine. A very few exceptionally fat and terminated(!) cylinders came from the Monserrat (Ichucollo) Mine, 12 kilometers east of Poopó.

Cylindrite ore was shipped to the Muldenhütten smelter in Germany for many years, until the Bolivians devised their own process for separating the tin, lead, antimony, and silver (0.5 weight percent) which comprise it. Unfortunately, separating this complex mixture is expensive. Large-scale exploitation of cylindrite has, thus, been carried out only erratically; the last significant lot came out in 1995, when a few hundred kilograms of cylindrite (albeit, only thin cylinders) were saved for collectors. Production has currently halted, although the cylindrite-bearing veins have not yet been exhausted.

Cylindrite is very rare on Poopó dumps, although good specimens could still be collected in the 1970s around mine buildings and collapsed miners' houses at the Trinacria Mine. Some of these specimens contained unusually lustrous cylindrite crystals in reddish brown, fine-grained sphalerite, making an attractive contrast.

Terminated cylindrite specimens have occasionally been available from other Bolivian mining districts: the Itos Mine in the city of Oruro (page 26) and the Siglo XX Mine in Llallagua (page 50). Cylindrite has also been reported from Carguaycollo, Colquechaca (page 63), and the defunct Porvenir Mine in Huanuni (page 18), but none of these localities has ever produced quantities approaching those produced by Poopó.

Franckeite, $Pb_6Sb_2FeSn_2S_{14}$, is another unusual tin sulfosalt that is common in this district, and a few other Bolivian localities but is rare elsewhere in the world. It was exploited for its high silver content (about 1 weight percent) as well as its tin. Franckeite is closely related both chemically and structurally to cylindrite, although it very rarely displays the cylindrical habit, occurring more usually as tabular, striated crystals or, most commonly, as flaky masses.

Teallite, $PbSnS_2$, is a tin-lead sulfide that can look superficially similar to franckeite but is much less widely distributed. A lot of franckeite has been sold to species collectors labeled "teallite." The original 1904 description of teallite was based on specimens found in an ore shipment of unknown locality, although later authors named the Monserrat Mine in the Poopó district as the type locality.

At the Monserrat Mine, teallite occurs in enough abundance to have been an ore mineral. It is sometimes recovered as pure masses but is more commonly mixed with wurtzite, sphalerite, and cassiterite. Small, lustrous, thin, platy crystals have also been found in Monserrat but are very rare. The thin crystals are so flexible that they can be rolled or folded just like aluminum foil!

The Antequera and Candelaria mines in the same district have also yielded teallite. But in recent years, a greater abundance of teallite specimens have come from the Carguaycollo Mine in Potosí department. The mine recently yielded a unique find of platy crystals, reaching several centimeters, associated with doubly-terminated quartz crystals.

The Candelaria Mine, about 8 kilometers south of Poopó, also produced the rare tin sulfide **herzenbergite** (SnS), named for German-Bolivian ore chemist Roberto Herzenberg (1885–1956). Herzenbergite is related to teallite and shares similar characteristics.

Other Minerals

Wurtzite, (Zn,Fe)S, is abundant in the Monserrat Mine, with pyramidal crystals reaching 1 centimeter wide. The best Bolivian "needle-tin" **cassiterite** specimens also come from Monserrat. This cassiterite variety forms rich aggregates of light yellow to almost colorless needles up to 2 centimeters long, with pyramidal terminations. Fibrous **zinkenite**, $Pb_9Sb_{22}S_{42}$, occurs in association with sphalerite in the Trinacria and Bolivar mines, while silver-bearing **jamesonite**, $Pb_4FeSb_6S_{14}$, is found with cylindrite in the San Francisco Mine. Very rare, cherry-red, radial aggregates of **kermesite**, Sb_2S_2O,, an oxidation product of of jamesonite, occur at a number of mines in the district.

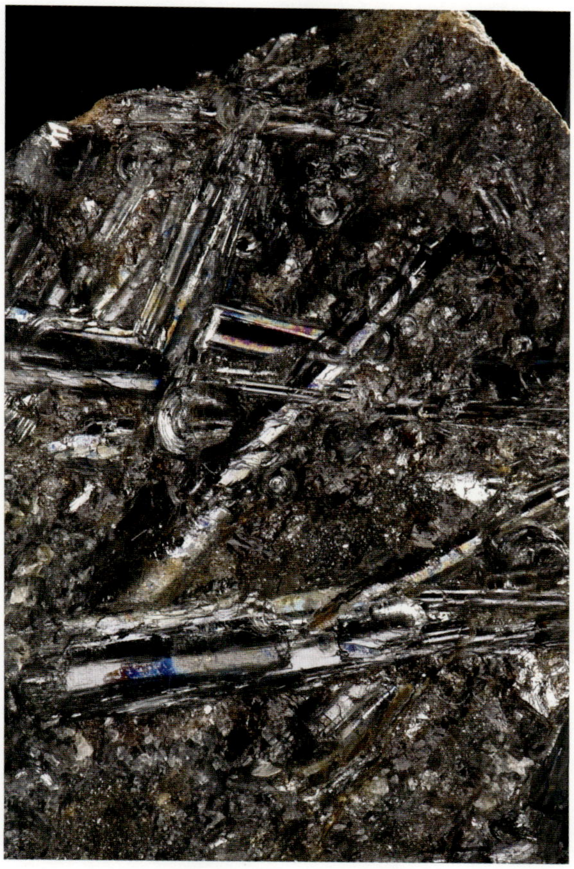

Left: *Cylindrite from the Santa Cruz Mine, Poopó, Oruro department; the field of view is approximately 4 cm wide; Arkenstone specimen; Joe Budd photo*

In the 1940s and 50s, long before **vivianite**, $Fe_2^{+3}(PO_4)_2 \cdot H_2O$, was discovered at the Morococala (page 32) and Huanuni, the Poopó district was the world's premier vivianite source. It commonly forms radial aggregates of acicular crystals in the Monserrat Mine. A 1948 find yielded blue-green tabular crystals from large vugs at the 240-meter-level. The largest crystals measure 20 (!) centimeters. One kilometer up the river that runs through town, past the women washing clothes in the geothermally heated waters, violet apatite crystals have been found on the dumps along the riverbanks.

Silver and Tin Minerals from the

Robert Cook *is Professor Emeritus in the Department of Geology and Geography at Auburn University, Alabama, USA where he was Department Head for 22 years. He is author of approximately 200 publications relating to mineralogy and economic geology and is an Executive Editor of* Rocks & Minerals *magazine.*

Andorite on stannite from the San José Mine in Oruro, Cercado province, Oruro department Bolivia; 3.6 cm tall; Mike and Sally Bergmann specimen; Jeff Scovil photo

San José and Itos Mines in Oruro

Oruro, a rather bleak city of approximately 200,000 inhabitants, sits at the base of a series of volcanic hills that rise 200 to 300 meters above the surrounding Altiplano, which, itself, is at an altitude of 3,700 meters. The city is some 200 kilometers southeast of La Paz and is the center of government for the department of Oruro. Its mines, of which the San José and Itos are the best known, have been the source of large quantities of silver and tin. Although not exhausted, they are today worked only by miner cooperatives similar to those described for Llallagua. The mines here are well-known for fine specimens of andorite, zinkenite, franckeite, stannite, boulangerite, and cassiterite (variety needle tin).

Left: One of the stained glass windows in the Santuario de la Virgen del Socavón church in Oruro; the church hosts a mining museum. Frank de Wit photo (2009)

Below: The tailings of the Itos Mine near Oruro Frank de Wit photo (2009)

The early history of the Oruro area and its deposits has become clouded by time. Tradition states that the Incas mined silver here and that the Spaniards ultimately found the deposits, taking them over in about 1595. Some accounts say that the city was founded on November 1, 1605 and originally named Real Villa de Don Felipe de Austria in honor of the then Spanish monarch Felipe III. The name was later changed to Oruro for the Urus region in which it lies. In the seventeenth century, Oruro was the largest city in what was then known as Alto Peru and its importance as a silver producer equaled that of Potosí far to the south. Oruro's mines were quite active until the War of Independence in 1780, at which time there was a severe decline in production. Mining was revived about 1845, driven by high silver prices and improved tin processing techniques.

In 1906, a modern mill was constructed at Machacamarca, about 30 kilometers south of Oruro, to treat San José and other Oruro ores. Following the nationalization of Bolivia's mines in 1952, the state-run corporation known as COMIBOL built a flotation plant near the site of the Itos Mine. It and the nearby San José Mine became the dominant producers, operating at depths approaching 500 meters. This mill shut down in 1991 after having produced almost 2 million tonnes of tailings. Significant mining at Oruro ceased with the closure of the San José Mine in August 1992. An unsuccessful attempt to leach the Itos tailings for remaining silver and other metals was made about 2000.

In the early 2000s, the San José Mine was reopened by a group of four cooperatives, which have the hoists working again and are exploiting small amounts of ore from the deepest levels. Tours into those deep levels are reportedly available and surprisingly good specimens are appearing once again.

For those not wishing to venture into an active mine, some of the earliest mine workings can be seen from inside Santuario de la Virgen del Socavón church, where the workings were found and preserved during expansion of the church. An excellent collection of Bolivian minerals, including a part of the Herzenberg collection, is on exhibit in Universidad Técnica de Oruro.

Geology

The ore deposits at Oruro are typical Andean subvolcanic polymetallic vein systems. The veins are intimately associated with intensely altered Tertiary rhyolite to quartz-latite porphyry stocks near their contact with shales of the Uncía formation. In the deeper levels of the San José Mine, the Uncía shales are intensely silicified and tourmalinized

Upper: Andorite San José Mine; 7.9 cm tall; Canadian Museum of Nature specimen (ex Bill Pinch); Michael Bainbridge photo

Lower: Augelite San José Mine; 5.5 cm wide; Kosnar collection and photo

adjacent to veins. Mining has been conducted to about 500 meters below surface, and diamond drilling has shown that veins extend at least to 700 meters. As with many such deposits, there is vertical zoning with respect to mineralogy and geochemistry. Ores are dominated by silver and tin above 180 meters and by silver and base metals below a depth of about 220 meters. Oxidation reached depths of up to 150 meters in the wider veins.

Paragenetic studies suggest two stages of mineralization. The first was relatively high-temperature, emplaced at shallow depth and of short duration, producing an assemblage of cassiterite, pyrite, quartz, and arsenopyrite. A second, lower-temperature stage resulted in the deposition of silver sulfosalts, jamesonite, tetrahedrite, andorite, franckeite, boulangerite, zinkenite, galena, chalcopyrite, and pyrite. The second-stage veins are commonly vuggy, giving them good specimen potential. Some veins are particularly silver-rich down to at least the 350-meter-level, and andorite was found in local abundance to the deepest levels in the San José Mine. The veins, several dozen of which were important producers, can be placed into one of three groups: a north-south group containing most of the larger veins of the San José Mine, a northeast trending group, and an east-west group.

Minerals

Mineral specimens have always been difficult to acquire in Oruro. Historically, there has been a relatively large technical population in terms of metallurgists, geologists, and engineers; and a few of them have been active collectors who, by virtue of their constant presence, acquired the best material. This was certainly the case in the 1970s and 1980s, when good specimens in any quantity were hard to come by if one was an outsider. However, an appreciation of specimens and their values has slowly developed among the mining populace, and small lots of good material are occasionally available. Collecting is a challenge but is certainly possible. Some dumps contain good pyrite specimens, and bright cubes and octahedra to several centimeters on matrix are occasionally found. Ore piles near mine entrances also can be the source of franckeite, teallite, boulangerite, and andorite. Stannite could be collected on dumps near the old Itos Mine in years past. Those species for which Oruro is best known or for which specimens have been recently available are briefly described below.

Andorite-IV/VI – $Ag_{15}Pb_{18}Sb_{47}S_{96}/PbAgSb_3S_6$

The finest andorite specimens known have been found in the Itos and San José mines. According to Moëlo et al. (1989), they are intergrowths of andorite IV with andorite VI. Vuggy masses of andorite containing relatively lustrous, dark gray crystals, which individually reach 1 centimeter, are occasionally available, and crystals up to 5 centimeters were reported in old literature. Andorite in crystals to 3 centimeters and displaying complex forms were described from the upper levels of the Purisima and Bronce veins of the San José Mine in 1897. Small andorite crystals in quartz crystal-lined quartz vugs have been found on the Itos and San José dumps.

Over the past several years, fine andorite specimens have appeared in crystals to 6 centimeters. In addition, two magnificent andorite specimens, one with crystals to 10 centimeters, were found recently by cooperative miners. Both specimens reportedly went to European museums.

Augelite – $Al_2(PO_4)(OH)_3$

Augelite in good specimens consisting of sharply formed, colorless to milky white, tabular to blocky crystals to 8 millimeters, densely grouped on matrix, have been found on a number of dumps in the general San José-Itos area.

Boulangerite – $Pb_5Sb_4S_{11}$

Boulangerite is widely distributed in the Oruro mines, and it is often impossible to distinguish, without X-ray analysis, thick boulangerite prisms from similar appearing zinkenite. Felted masses of boulangerite needles are common in vuggy ore from deeper levels of the San José Mine and are similarly impossible to visually distinguish from jamesonite. Boulangerite has been found associated with pyrargyrite at the Itos Mine.

Bournonite – $PbCuSbS_3$

Although virtually unobtainable today, magnificent bournonite specimens were well documented from the Bronce vein in the late 1800s, and single and groups of untwinned bournonite crystals were found in the Grande vein in the Colorada section of the San Jose Mine in the 1940s.

Cassiterite – SnO_2

Although certainly not noted for good cassiterite specimens in the sense of Viloco (Araca) or Huanuni, the upper levels of the Itos and San José mines have produced unusual "needle tin" specimens as sprays of medium-brown crystals in vuggy pyrite.

Chalcostibite – $CuSbS_2$

Chalcostibite occurs in Oruro ores, occasionally in fine specimens. The species is best known as bright, steel gray, tabular or blade-shaped crystals up to 1 centimeter in length that are associated with andorite, stannite, quartz, and pyrite in massive tetrahedrite.

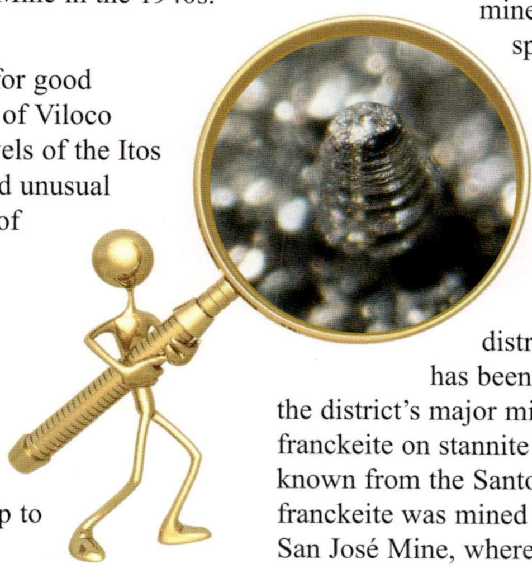

Cylindrite – $(Pb,Sn^{2+})_8Sb_4Fe_2Sn^{5+}_4S_{27}$

Although certainly an uncommon mineral at Oruro, good cylindrite specimens are occasionally encountered. An unusual recent find included free-standing terminated cylindrites in vugs with quartz and franckeite.

Franckeite – $Pb_5Sn_3Sb_2S_{14}$

Franckeite is widely distributed in Oruro veins, and it has been described in ores from all of the district's major mines. Interesting specimens of franckeite on stannite and zinkenite crystals are known from the Santo Cristo Mine, and abundant franckeite was mined from the Grande vein in the San José Mine, where it occurs with plagionite and

Above left: *A mass of boulangerite-jamsonite-zinkenite that was just removed from a pocket in the San José Mine Frank de Wit photo (2009)*

Above right: *Frankeite from Oruro; 10 cm tall; Joan Massagué specimen; Jesse La Plante photo*

Inset: *Cylindrite from the Itos Mine; 0.5 mm tall; John A. Jaszczak collection and photo*

***Right**: Plagionite; 3.3 cm tall; Joan Massagué collection; Joe Budd photo*

***Below**: Stannite and andorite; 5.8 cm wide; Fabre Minerals specimen; Jeff Scovil photo*

***Inset right**: Franckeite from the San José Mine; 6 mm long; John A. Jaszczak collection and photo*

galena. Coarse franckeite, with individual tabular crystals to 2 centimeters lying in a matrix of fine-grained teallite, are common in the deeper levels of the San José Mine. Ball-like franckeite aggregates to 2 centimeters with filiform zinkenite have been available from the San José Mine in recent years.

Gold – Au

Interestingly, Coogan Gold Company offered Bolivian gold specimens for sale in 2001. They were collected on several mine dumps on the periphery of the city of Oruro. The small specimens, which likely originated in the Iroco Mine vicinity, contain sheetlike masses of gold in matrix and were found using a metal detector.

Jamesonite – $Pb_4FeSb_6S_{14}$

Some of the world's finest jamesonite occurs in the San José Mine. Unfortunately, it is essentially impossible to distinguish between it and similar-appearing zinkenite or boulangerite without X-ray diffraction analysis; thus unavoidable mislabeling likely exists. Unusually thick prismatic crystals to 8 centimeters long have been collected recently.

Metastibnite – Sb_2S_3

Metastibnite specimens that are unusually good for the species were found recently in the San José Mine. The mineral occurs in thick botryoidal crusts with a strong metallic luster and local golden tarnish, quite unlike the more typical thin red films seen from other localities. The mineral is distinguishable by its red streak.

Miargyrite – $AgSbS_2$

Fine miargyrite specimens with well-developed crystals reaching 1 centimeter were at one time found in the Bronce vein, and excellent miargyrite specimens attributed to the Itos Mine are seen in old collections.

Plagionite – $Pb_5Sb_8S_{17}$

Some of the finest plagionite specimens known have been found in the Purisima vein. Plagionite occurs as coarse tabular crystals on meneghinite and as up to 1-centimeter-wide crystals associated with galena and "needle tin." The same vein produced plagionite crystals to 5 millimeters that are intimately associated with ball-like aggregates of franckeite.

***Right**: Descending into the San José Mine, miners are "protected" by swollen timbers, which are compressed by the country rock. Frank de Wit photo (2009)*

***Inset right**: Rhodostannite The largest crystal is 0.5 mm wide. David K. Joyce specimen and photo*

Rhodostannite – $Cu_2FeSn_3S_8$

The finest known specimens of this rare mineral were found a few years ago during the re-examination of old andorite specimens. The rhodostannite occurred as sharp pseudo-octahedral crystals to about 1 millimeter.

Stannite – Cu_2FeSnS_4

Stannite is well known from Oruro; its axial ratios were first described from crystals collected in the Bronce vein there. The mineral is typically seen in groups of small, somewhat tabular, dark gray crystals associated with pyrite, arsenopyrite, zinkenite, andorite, and quartz. Rarely, outstanding specimens are available. In 2004, platelike well-formed, complexly twinned stannite crystals to 2.1 centimeters grouped with sprays of zinkenite prisms (some of which were later revealed to be jamesonite) to 3.5 centimeters were found in the San José Mine. The next year stannite crystals to 5 centimeters were found in the same area. Other finds include ball-like aggregates of stannite crystals to 5 centimeters across that occur on matrix or as floaters in vugs.

Teallite – $PbSnS_2$

Although not nearly as abundant as franckeite, teallite is nevertheless an important mineral in Oruro ores. Probably the finest crystals known, which occur as square plates exceeding 1 centimeter in width, have been described from the San José Mine.

Valentinite – Sb_2O_3

Exceptional 2-centimeter valentinite crystals, some of the finest known, were found in the 1950s and again in the mid-1970s in the San José Mine. These crystals are of a light tan color and should not be confused with the butterscotch-colored crystals available from the Tatasí Mine, in south-central Potosí.

Wavellite – $Al_3(PO_4)_2(OH,F)_3 \cdot 5H_2O$

Although not noted for phosphate species, Oruro veins have produced excellent wavellite specimens. Colorless crystals up to 5 millimeters wide form groups on quartz and occasionally stannite and were found in the upper levels of the Bronce, Purisima, and San Luis veins of the San Jose Mine. Microcrystalline wavellite can still be found on some mine dumps.

Zinkenite – $Pb_9Sb_{22}S_{42}$

Zinkenite is one of the most widely distributed sulfosalts at Oruro. It occurs in most major veins and is seen in vuggy ore as toothpick-like smoothly lustrous prisms to 5 centimeters long that are often associated with andorite and stannite. Solid vuggy aggregates of zinkenite crystals to 8 centimeters across were once relatively common in San José ore. However, distinction between the similarly fibrous to prismatic minerals jamesonite and boulangerite is almost impossible without careful X-ray analysis; so specimens labeled as any one of these three species are suspect. In fact, X-ray diffraction analysis of some supposed zinkenite crystals of exceptional size and quality associated with recent finds of andorite and stannite, show that at least some of this "zinkenite" is jamesonite.

***Left**: Acicular zinkenite (analyzed) on twinned stannite crystals found in the San José Mine in October 2004. The overall specimen is 4 cm tall. Fabre Minerals specimen and photo*

***Inset left**: Teallite from the San José Mine; the largest crystal is 0.8 mm long; Dan Behnke collection and photo*

***Above:** A miner's helmet welcomes visitors entering Oruro from the west. Frank de Wit photo (2009)*

And others ...

Other species known to occur in collectible specimens within the Oruro district include galena, cerussite, semseyite, tetrahedrite, wurtzite, stibnite, pyrite, diaphorite, vivianite, apatite, orthoclase, and a variety of post-mining secondary minerals.

The Morococala Mine

by ***Jaroslav Hyršl*** *and* ***Alfredo Petrov***

Upper*: Vivianite on pyrite from Morococala, Dalence province, Oruro department; 3.5 cm tall; Dave Bunk Minerals specimen (ex Harold Michel collection); Jesse La Plante photo*

Morococala is a relatively small cooperative mine, about two hours east of Oruro, over a rocky dirt road. Before vivianite was discovered at this 4,300-meter locality in 1981, Morococala was just a tin mine and was of no importance to specimen collectors. The vivianite discovery brought with it a 12-year-long economic boom for some of the miners. Unfortunately, the best vivianite-producing levels flooded in the early 1990s. With a stack of unpaid electric bills and lacking the capital for pumping, the cooperative fell on hard times. Membership during the last few years has dropped from 300 to 70 miners.

The flooding of Morococala was interesting, as unlike most underground mines, Morococala did not need to be pumped and was assumed to have a natural drainage system. Two kilometers away (and downhill) is the larger Santa Fé tin and silver mine, operated by COMIBOL. The Santa Fé, like most underground mines, needed to be pumped. In the early 1990s when COMIBOL abandoned the Santa Fé, water began rising in Morococala at the rate of 2.5 centimeters per day, which, after several years, annihilated the lower workings.

Lower*: A 17 cm wide vivianite on a matrix encrusted with the drusy pyrite that is characteristic of specimens from Morococala. Carnegie Museum of Natural History specimen; Harold and Erica Van Pelt photo*

Morococala had been dry because the pumping of the deeper Santa Fé Mine produced a cone of depression more than two kilometers in diameter that encompassed the Morococala Mine. When the Santa Fé Mine closed and began to fill with water, the bottom of the cone of depression became more and more shallow, and the water level at Morococala rose and flooded the mine.

Geology

Morococala veins are emplaced in somewhat metamorphosed sediments. The presence of contact metamorphic minerals in the sandstone—andalusite, sillimanite, topaz, tourmaline, apatite—indicate that an igneous intrusion is probably at shallow depth, although it is not exposed anywhere. Unmineralized Pliocene lavas form a plateau over much of the surrounding area; Morococala and several neighboring tin-silver mines were discovered principally where erosional "windows" revealed the older mineralized rocks. There are probably, thus, other tin-silver deposits still hidden under the lavas.

Minerals

Vivianite from Morococala doesn't reach the meter lengths of crystals from Anloua in Cameroon, but they are still among the largest and most beautiful in the world. The first specimens reached the market in the 1970s although these were usually labeled "Santa Fé" for the nearby mine and the Santa Fé COMIBOL office, which served the district. Morococala vivianite crystals exceed 30 centimeters in length, and many are almost transparent. The best crystals came from the deepest levels of the mine, especially the 280-meter-level. Unfortunately, these deeper parts of the mine are now flooded, and chances of finding more specimens are slim.

Morococala's vivianite occurs on a matrix characteristically covered with iridescent, drusy pyrite microcrystals, which easily distinguish them from the vivianite from other Bolivian localities. Dark red-brown brittle amorphous **hisingerite** encrusts many Morococala vivianite crystals, especially those from the 211-meter-level. The crust is unfortunately common and diminishes the value of many otherwise beautiful specimens, although it can be removed in part with a water gun.

While many specimens are labeled otherwise, vivianite from Morococala never occurs in association with ludlamite. Specimens occurring with blocky ludlamite crystals are from Colavi; those with more pointed ludlamite crystals are from Huanuni. Similarly, groups of ludlamite NOT associated with vivianite are generally from Colavi or Huanuni and not from Morococala.

Phosphophyllite is a surprising find from this mine: On one occasion, the authors found rare radial "suns" of thin prismatic blue-green phosphophyllite crystals up to 7 millimeters long on large, dull (probably etched) vivianite crystals. We also found it as inclusions in large vivianite crystals.

The dumps still yield microcrystals of cassiterite, pyrite, sphalerite, and not much else. Nearby, the San Luis antimony mine has produced some fine **stibnite** crystals.

Above: *The upper half of the "glory hole" at Morococala Alfredo Petrov photo (1999)*

Left: *Stibnite and marcasite from the San Luis antimony mine, not far from Morococala; 19.5 cm tall; Bob Byers collection (ex Steve Neely); Jeff Scovil photo*

Left: *Ferberite from Tasna; 9.4 cm wide*
Christian Weise specimen; Jeff Scovil photo

Middle left: *Magnetite from Cerro Huañaquino*
5.8 cm tall; Kosnar collection
Brian Kosnar photo

Middle: *Sulfur from El Desierto; 6 cm tall*
Rock Currier collection and photo

Middle right: *Bournonite from Colquechaca*
4.7 cm wide; Collector's Edge Inc. specimen
(ex Philadelphia Academy); Jeff Scovil photo

Lower left: *Wavellite from Llallagua; 5.2 cm tall;*
Marty Zinn collection; Jeff Scovil photo

Lower right: *Phosphophyllite from Cerro Rico*
8 cm wide; Jim and Gail Spann collection
Jesse La Plante photo

Potosí Department

*Map by **Günther Neumeier***

The Minerals of Silver-Rich

Jaroslav Hyršl *is a Czech mineralogist, gemologist, and mineral dealer. He specializes in the minerals of Bolivia and Peru.*

Coauthor ***Alfredo Petrov*** *worked for many years as a geologist in Bolivia; Potosí was his favorite city. For Alfredo, after visiting this, the world's highest altitude city, the rest of life is downhill.*

Andy Seibel and Jerry Rosenthal spotted a hint of phosphophyllite in a quartz "boulder," from Cerro Rico, being offered by a dealer at the Sainte Marie show in 2006. Jerry bought the specimen, brought it home, and cleaned it, thereby exposing this rare phosphophyllite twin on quartz. The specimen is 5.5 cm tall. Jerry Rosenthal collection; Jeff Scovil photo

Cerro de Potosí aka Cerro Rico

History

Potosí might well be the richest silver deposit not only in Bolivia but in the world. The time-honored tale about the discovery of the deposit has it that in 1544 an Indian named Diego Hualca stopped on Cerro de Potosí while searching for his lost llamas. Diego built a fire, which melted the surrounding rocks and produced liquid silver. He told his friend Huanca about the find, and the two began to exploit it. They soon started to quarrel over the profits, and Huanca told the Spanish about the find. Naturally, the Spanish took the silver for themselves. Cerro de Potosí is widely and aptly known today as "Cerro Rico," meaning rich mountain.

The city of Potosí, at an altitude of 4,090 meters above sea level, was founded in 1545, and thousands of Indian slaves were sent to mine the precious white metal. It did not take much capital for a Spaniard to start a silver mine. One had only to register a claim, then ask the local authorities to hand over, at no cost, an appropriate number of Indian slaves. It is estimated that during the Colonial period (1545–1825), about eight million Indians were killed in Bolivian mines, victims of accidents, starvation, or silicosis! Consequently, large portions of the southern Altiplano were depopulated. At least 3,000 adits existed. Each mine had its own ovens for smelting ores, and more than 6,000 small fires lit Cerro Rico at night.

Potosí became immensely rich and was soon the largest city in the Western hemisphere, with about 250,000 inhabitants and 36 cathedrals. Big water reservoirs were built in the neighboring Kari-Kari Mountains, and in 1572, a mint was established. A later mint, the eighteenth century Casa de la Moneda, is now one of Bolivia's best museums; it boasts a number of interesting exhibits from the Colonial period, including some mineral specimens (the remains of the Gumucio collection)! Colonial silver coins were transported by caravans of llamas to Lima or Arica in Peru, then by ship to Spain, giving rise to piracy in the Caribbean.

One of the most important surviving references to early mining conditions and technological procedures in Potosí is the book *El Arte de los Metales* by Alvaro Alonso Barba (1640). The book was almost lost to the Spanish Inquisition in an early attempt to protect industrial secrets (Moore, 2008).

Mining in Potosí stopped during the chaos of the War of Independence in 1805 and did not resume until 1886, but even then it was mostly for a "new" metal: tin. In Colonial times, tin had been considered a nuisance by the silver smelters, although minor quantities of tin were produced for local uses such as retorts for distilling alcohol. Nowadays, silver production on Cerro Rico is increasing, although the value of the zinc, tin, and antimony production continues to exceed that of silver.

Ore processing at Potosí; Daniel Beams photo, 2008

After the Bolivian revolution of 1952, the Pailaviri (or Unificada) Mine, the largest mine on Cerro Rico, was nationalized and became a division of the state mining corporation COMIBOL. Like state-run enterprises everywhere, COMIBOL was inefficient and corrupt, and under its control, the operation was unprofitable. The State then gave the mines to the miners, who still run them as cooperatives, albeit primitively, without capital or technical support. Currently, between 5,000 and 15,000 miners work in and on Cerro Rico; the number of miners is a function of the world price of the various metals produced there.

Geology

The main rocks in the Potosí area are slightly metamorphosed Ordovician shales and minor amounts of sandstones overlain by volcanic tuffs, which are of probably mid-Tertiary age. A few kilometers east of Potosí, the Kari Kari diorite porphyry batholith is exposed. The most important volcanic rock in the region is the mushroom-shaped Potosí stock, which forms much of Cerro Rico and was intruded about 30 million years ago. It occupies a 1,700 by 1,200 meter area on the surface but shrinks to a diameter of only 100 meters in the

***Above:** Rare doubly terminated, twinned phosphophyllite crystal on matrix from the Unificada Mine 6.4 cm tall; Steve and Clara Smale collection; Jeff Scovil photo*

deepest levels of the mine. The original rock was probably dacite or quartz latite, but it is so highly altered that it difficult to identify definitively. Silicification prevails in the upper part of the Potosí stock (with up to 95.6 percent silica!) while while alteration of host rock to white micas is more prominent at depth. Macroscopically, the rock is characterized by complete dissolution of the original feldspars. The vugs that were left behind after the feldspar crystals disappeared are hollow or on rare occasions are filled with quartz, phyllosilicates, hematite, jarosite, and other secondary minerals, sometimes as interesting microdruses.

The mineralized veins on Cerro Rico mainly cut the Potosí stock and less frequently, the surrounding volcanics or sediments. They trend from north-south to northeast-southwest. Turneaure (1960) mentions 113 veins in the central vein system, but a sheeted zone, with a maximum width of 175 meters, in the center of the stock, is composed of hundreds of parallel veinlets, 1 to 5 centimeters in width. The average length of the veins in the central vein system is about 125 meters. The veins are lenslike, ranging in width from 2 to 150 centimeters. Veins formed in two main phases: the first generation is characterized by cassiterite and pyrite, with minor ferberite and bismuthinite at depth; younger veins contain more varied minerals, with stannite, tetrahedrite, sphalerite, and pyrargyrite prevailing. The veins contain vertical zones, with rich silver ores and cassiterite in the upper parts, the latter changing to stannite at depth. Horizontal zoning is poorly defined, but veins on neighboring Cerro Chico contain mainly sphalerite, wurtzite, galena, tetrahedrite, and boulangerite, but no cassiterite. Potosí geologists have recently observed that the veins contain predominately silver ore in the higher reaches of the stock and are more sphalerite-rich at depth. Cassiterite is more pronounced in the center of the stock, toward the bottom part of the sheeted zone. The origin of the Potosí deposit is associated with mid-Tertiary volcanic activity. It formed at shallow depth, under high temperature (up to 500 degrees C) and relatively low pressure.

The oxidation zone occupies the upper third of Cerro Rico. Some parts were reputedly very rich, composed mainly of chlorargyrite, argentite, and native silver, but its exact character is now unknown due to the intensive Colonial mining efforts-hardly any ore minerals remain. Cassiterite from limonite-jarosite veins, however, is still mined from the upper portions of Cerro Rico.

Although extremely rich silver veins exist in Cerro Rico, most of Spanish Colonial production was from very poor ore, which would be impossible to mine economically today. The profitability of mining that ore was made possible by the large-scale of the workings and the widespread use of Indian slaves (Alonzo de Barba, 1640).

Phosphophyllite

Potosí is world famous among mineral collectors for superb phosphophyllite crystals. These often-twinned, transparent pale blue-green crystals to 15 centimeters are without doubt the very finest of the species. Larger twins are among the most expensive mineral specimens on the collector market and have been called the Holy Grail of mineral collecting; so much so that veteran South American dealer Terry "Skip" Szenics (szenicsite) recently quipped, "a phosphophyllite crystal is akin to sex: even when it's bad, it's damn good!"

An Argentine mineralogist showed a phosphophyllite specimen from Cerro Rico to Fred Pough in the 1930s, but until the German ore chemist Fritz Berndt (berndtite) identified them as phosphophyllite in 1957, no one in Bolivia attached much value to the crystals; so the miners' children hammered them into pretty cleavages and played with them. The best specimens were found in the Krause vein in the late 1950s and 1960s. The miner who found the top specimen—a 14-centimeter twin on a sulfide matrix probably worth seven figures—allegedly traded it for a refrigerator.

The specimens found in more recent years have been lower in quality and quantity; nevertheless,

A 9.78 carat phosphophyllite cut by master lapidary Art Grant; Private collection; Tino Hammid photo

there has been a trickle of new specimens, mostly from narrow inch-wide sphalerite veins in which phosphophyllite rarely has the opportunity to form large, well-terminated crystals. Many of the recent malformed and broken crystals are sold as cutting rough. The "neocolonial" era, during which fabulous specimens could be traded for a refrigerator and smaller specimens for a couple of dollars or a bottle of beer, is long gone. Today's miners have access to the Internet. They study international dealers' websites and are well aware of what good specimens sell for in wealthier countries, and they expect to be paid accordingly—at least a Jeep in trade for a good specimen!

Silver Minerals

Although phosphophyllite is the first mineral that comes to mind when collectors talk about Potosí, its silver species are also highly sought. **Native silver** (Ag) was supposedly common in Colonial times, but few specimens have been preserved. Recently, however, a few silver wires up to about 2 centimeters long, growing on rock and covered with white clay, were collected in the highest part of Cerro Rico.

Pyrargyrite (Ag_3SbS_3, trigonal) is one of the most important silver minerals in Cerro Rico. Very rich occurrences of massive pyrargyrite were found in 1994 in the Potosí vein in both a cooperative mine and on a lower level of the same vein in the Pailaviri Mine. Cavities in this vein very rarely contain **pyrostilpnite** (Ag_3SbS_3, monoclinic) crystals up to a few millimeters long in association with older valentinite crystals. Beautiful, complex, translucent, dark red pyrargyrite crystals up to 1 centimeter in diameter were found in a vein with apatite crystals in January 2003. Massive **miargyrite** ($AgSbS_2$) was commonly found in a superficial part of the Tajopolo vein, but in 1995, lustrous crystals up to about 5 millimeters and clustered in rich granular aggregates were found lining cavities of diaphorite and **barite**. Even larger miargyrite crystals have been reported recently.

Massive **andorite** in association with pyrargyrite has been mentioned in the literature. Columnar crystals about 3 millimeters long, sometimes as hollow tubes(!), were found by the authors in association with franckeite. **Diaphorite** ($Pb_2Ag_3Sb_3S_8$) was first found in Cerro Rico in 1995. It forms striated crystals up to 3 millimeters on massive pyrargyrite. It is less lustrous and more rounded than the accompanying miargyrite. **Chlorargyrite** used to be one of the most important minerals of the oxidation zone, but specimens are very rare now.

Above right: *Phosphophyllite; 6.8 cm tall; Houston Museum of Natural Science specimen; Jeff Scovil photo*

Insets

Top: *Pyrostilpnite; 2.7 mm long, Bill and Carol Smith collection; Dan Behnke photo*

Middle: *Miargyrite crystals, the largest is 2.5 mm wide Bill and Carol Smith collection; Dan Behnke photo*

Bottom: *Pyrostilpnite (red), diaphorite (grey) with calcite (beige); the field of view is about 1 cm; Jaroslav Hyršl collection and photo*

Dr. Fritz Berndt and the True Type Locality of His Namesake Mineral

by Alfredo Petrov

During the second World War and for the first two decades after the war, foreign geologists and mine engineers from the United States, Germany, Poland, Yugoslavia, and even New Zealand filled most of the technical positions in the Bolivian mining industry. Several of these men were also avid mineral collectors. Among these collector/mining professionals, Mark Chance Bandy was an American who worked for Simón Patiño at the Siglo XX Mine; Friedrich (Federico) Ahlfeld, a German geologist, worked for the Bolivian government; Robert (Roberto) Herzenberg, an ethnic German from Latvia, was the chief ore chemist for the German-Jewish tin baron Moritz (Mauricio) Hochschild; Fritz Berndt, another German chemist, also worked for Hochschild; Hans Block, a German prospector and mine owner, also held positions in the government. The collector group was rounded out by Polish mine engineer Pyotr Zubricky. All of these men assembled important collections of Bolivian minerals; the first four have rare species named for them: herzenbergite, bandylite, ahlfeldite, and berndtite. Berndtite is the subject of this story, as an incorrect type locality is often erroneously associated with it.

During the 1952 revolution, with the overthrow of the tin barons and nationalization of the mines, Moritz Hochschild escaped Bolivia. Assuming that all westerners in Bolivia would be terrified of the impending socialist revolutionary government and would be anxious to leave, Hochschild offered his German employees Herzenberg and Berndt, and even nonemployee Ahlfeld, jobs at Hochschild's other mines, in Peru. Surprisingly, all three men enjoyed the lifestyle and mineral collecting in Bolivia; so they elected to stay and try their luck with the new government. Robert Herzenberg became chief chemist at COMIBOL's ore analysis laboratory in the city of Oruro. Fritz Berndt took over this position after Herzenberg's premature death while traveling in Chile.

One day, while studying a polished ore section of stannite that was partially altered to fine-grained cassiterite, Berndt noticed two unidentifiable semi-metallic mineral phases that were apparently intermediate products of stannite's decomposition to cassiterite. One was pale brown, with intense yellow-orange internal reflections and the composition SnS_2. The other was grey, with orange-brown internal reflections and the composition Sn_2S_3. Enlisting help to describe these phases, Berndt sent the entire sample to German mineralogist Günter H. Moh.

In 1964, Moh and Berndt published their findings in the paper, *Two New Natural Tin Sulphides, Sn_2S_3 and SnS_2*, stating that the specimen came from "a zone of secondary enrichment or oxidation at Cerro de Potosí, Bolivia." They also state that these phases "are unnamed in this preliminary description pending the completion of further studies." In fact, the two new minerals remained unnamed for two years, until Günter Moh (1966) published, *Das Binäre System Zinn-Schwefel und seine Minerale*. In this paper, Moh named SnS_2 "berndtite," after the discoverer Dr. Fritz Berndt, and he named the Sn_2S_3 phase "ottemannite," in honor of Heidelberg mineralogist Joachim Ottemann (born 1914).

Mail service to Bolivia was slow in those days, and considerable time had passed before Dr. Berndt received a copy of the article and found out that one of the new minerals had been named for him. Berndt was embarrassed, because he was one of the authors of the original paper, and he did not want anyone to think he had had anything to do with naming a species after himself!

In spite of the fact that Cerro Rico is clearly the type locality of berndtite and ottemannite, various compilations of mineral data have given an incorrect type locality, causing confusion among collectors. The 2008 edition of *Fleischer's Glossary of Mineral Species*, for example, gives "Mina Teresa" [*sic*] (actually the María Teresa Mine) near Huari, Oruro department, as the type locality for berndtite and states that the type locality of ottemannite is "unknown"! Other sources have given the type locality of berndtite as the Panasqueira Mine in Portugal, an error that might have come about eight years later when Clark (1972) described the new hexagonal berndtite polytype, berndtite-C27, from Panasqueira. The original berndtite, from Bolivia, is differentiated as the berndtite-6C or berndtite-2T polytype. Incidentally, polytypes are no longer recognized as separate mineral species, so the species name is simply "berndtite."

Among the subclass of mineral collectors known as "species collectors" in the United States and "systematik collectors" in Europe—those who want one of every known species, regardless of aesthetics—berndtite is one of the most difficult species to obtain. No one has a piece from the type locality, not even Dr. Berndt. As for the alleged specimens from other localities such as the María Teresa Mine or the burning coal fumaroles of Forestville, Pennsylvania, the "berndtite" in them is often, as Vandal King likes to put it, "so rare it is not even on the specimen."

For those who may be curious about the disposition of the Bolivian collections mentioned above, the Bandy collection is now at the Los Angeles County Museum of Natural History in California. The Herzenberg collection went to the Universidad Técnica de Oruro, where it is on display, although the more valuable specimens seem to have been pillaged. The top specimens from Ahlfeld's collection were acquired by a private collector in the United States; the remainder is in a private collection in Bolivia. Most of Fritz Berndt's collection is conserved in Prague by Jaroslav Hyršl. The Block collection is in private hands in Bolivia. Pyotr Zubricky's excellent collection is the only one that can no longer be accounted for, although it is rumored to have been buried in the soil under the floor of a humble hut in a remote mining camp during the 1952 revolution, where it apparently remains!

Postscript: Dr. Fritz Berndt, now in his mid-90s, enjoys life in Cochabamba. He lives in a lovely home originally built by the late Dr. Friedrich Ahlfeld. This is perhaps the only example of a house occupied consecutively by two unrelated people who have both had minerals named for them!

Right: *Wire silver from Cerro Rico; 3.5 cm wide; Jaroslav Hyršl collection and photo*

Left: *Arsenopyrite with pyrite and quartz from Cerro Rico; the crystal group is 2.5 cm wide; Kosnar collection; Brian Kosnar photo*

Other Species

Stibnite is very rare in Potosí, but several nice druses of needlelike crystals were found in 1995. Collectors can find sharp, tabular, red-brown hexagonal **wurtzite** with **sphalerite** and **tetrahedrite** group minerals on the dumps and in vugs. **Arsenopyrite** is common in lower parts of the veins, where well-developed crystals up to 1 centimeter were recently found in association with stannite crystals. **Tetrahedrite** is likewise common in Potosí, occurring as euhedral crystals up to about 1 centimeter. These minerals are often silver-rich, but there are few analyses confirming the exact species present. **Bournonite** occurs as well-developed frosted "cogwheels," which can exceed 1 centimeter. These are found on a tetrahedrite matrix.

Stannite crystals were identified from Cerro Rico as early as the nineteenth century, but they cannot compete with the specimens from Oruro or Llallagua. This mineral is more abundant in deeper parts of the deposit. Druses of lustrous stannite, which can reach 5 millimeters, occur with arsenopyrite and quartz and in some cases with a younger generation of wavellite spheres.

Berndtite and **ottemannite** are very rare tin sulfides for which Cerro Rico is the type locality (see facing page). Franckeite was found here in 1995 in druses filled with small flat crystals in association with andorite. Feltlike **jamesonite** occurs on some **franckeite** specimens.

A spectacular first find of **valentinite** from Cerro Rico was made in 1995. Valentinite forms excellent transparent, lustrous crystals up to 1.5 centimeters long in vugs of massive pyrargyrite and is accompanied by younger small **pyrostilpnite** crystals! Many other finds of valentinite crystals up to 1 centimeter have been made since. Most of these occur as small druses without matrix and are very similar to specimens from the Colavi Mine. Crystals from both mines exhibit an intense adamantine luster and are partly colorless and transparent and partly clouded by inclusions of black sulfide specks.

Small crystals of **apatite-(CaF)** were found with

Above: *Chlorargyrite from "Cerro de Potosí;" 4 cm tall; Jaroslav Hyršl collection (ex Berndt) and photo*

Left: *Stibnite from Cerro Rico; 4 cm wide; Jaroslav Hyršl collection and photo*

the best phosphophyllite crystals in the 1950s in the Krause vein. Lustrous white **apatite-(CaF)** crystals to 1 centimeter accompanied some of the **pyrargyrite** from the 2002 find in the Potosí vein.

Melanterite is the most common secondary mineral in old galleries and stopes. Beautiful, emerald green, transparent melanterite stalactites can reach lengths of 30 centimeters. Miners sometimes pluck these from the walls and roofs of the galleries to give visitors as souvenirs. Unfortunately, they are unstable and quickly dehydrate to a white powder when taken outside the mine.

Collecting Opportunities

The tourist industry is more developed in Potosí than in any other Bolivian city, and there are numerous travel agencies and hotels in the downtown area advertizing underground tours of Cerro Rico's mines (see facing page). Every day, dozens of European, American, Brazilian, Israeli and Japanese tourists go into the mountain, and the majority come back alive—even when the tour guides detonate a stick of dynamite underground for the visitors' amusement, something they often do.

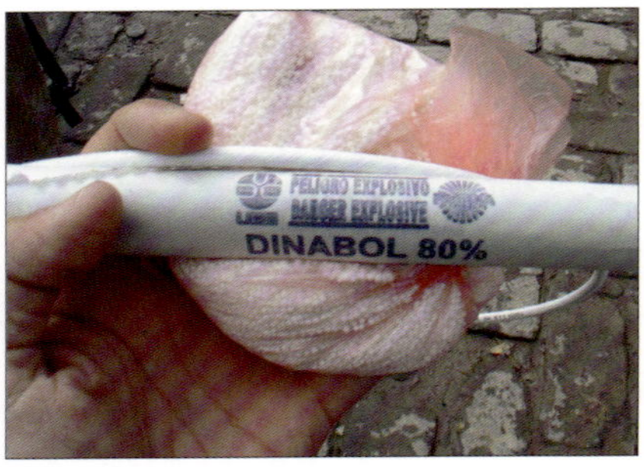

Explosives can be purchased in Potosí markets and are a welcome gift for a miner. Mark Pirlo photo

Most of the tourists know nothing about minerals and eagerly absorb all the nonsense told them by the guides. They might, for example, quote the numbers of miners who die from exposure to the long white "asbestos" fibers on the mine walls when, indeed, the material isn't asbestos at all; instead, the fibers are the water-soluble sulfates pickeringite and halotrichite. If a typically 3-hour-long tour of this nature appeals, by all means, go. Dress in old clothes, as you will get dirty, but no one is likely to object if you take your rock hammer along and stab at some of the veins in the walls.

If you prefer to do more serious collecting, there are no fences on the dumps and you can clamber all over them from the bottom of the mountain to the top. In fact, the surface of the mountain is barely exposed—it is covered from top to bottom by dump material. Naturally, the chances of one's finding a big phosphophyllite or pyrargyrite crystal on the dumps are about as likely as winning the lottery, but a lot of interesting micro material can easily be picked up on the dumps, and metallic ore minerals can be found on the ore piles stacked up by adit entrances. If you take specimens from the ore piles, you should of course give the miners or their wives and children something in return for them. Larger specimen material on the dumps includes pseudomorphs after feldspars and colorful, highly iridescent crusts of iron oxides known as "turgite."

Another and more exciting option for the adventurous, especially if you speak at least a bit of Spanish, is to organize your own underground tour and accompany a miner to his workface, where you can spend a whole shift digging in a vein. The miners who agree to allow you to do this will expect you to provide money for alcohol and coca leaves.

The "New" Bolivian Ecology

Over three centuries of Spanish Colonial control, the thousands of smelting ovens changed Bolivia's ecology. It takes ten tonnes of wood to make one tonne of charcoal, and two tonnes of charcoal to smelt one tonne of ore. So much charcoal was produced that many thousands of llamas were continuously occupied as endless trains carrying charcoal to Potosí. Naturally, it is impossible to calculate the exact statistics, but the charcoal used in Potosí alone (not counting all the other silver mining centers in Bolivia) represented something on the order of one million tonnes of vegetation destroyed every month.

Wild vicuñas grazing on the Altiplano; Daniel Beams photo, 2008

When the Spanish first arrived in the Andes, much of the Altiplano and Eastern Andes were covered with low forest: woody "thola" bushes up to about two meters high on the flat Altiplano and scrawny quewiña trees, with their papery red bark, on the steep mountain slopes up to almost 5,000 meters in elevation (the world's highest altitude tree species). A Franciscan monastery has a copy of an old letter that a conquistador wrote to the Spanish government complaining that the "forest" (probably tall thola bushes) between Oruro and Caracollo, 40 kilometers away, was so thick that it was difficult to pass through on horseback.

Today, not a single bush is visible between Oruro and Caracollo; instead, the view is of salty mudflats and grassland! The highlands of Bolivia were almost completely deforested and have never recovered, in large part because the sheep, goats, and pigs, introduced by the Spanish, eat any newly sprouting tree seeds.

Tour Potosí

Australian exploration and environmental geochemist ***Mark Pirlo*** *took a 2004 tour of an ore processing plant in Potosí then went underground for an organized tour of the Candelaria Mine. Here is some of what he saw.*

All of the photos on this page are by Mark Pirlo.

Cerro Rico

The mines of this rich mountain, at one point, made Potosí the richest city in South America

The Plant

Left: *Ore crushing*

Middle: *Froth flotation*

Right: *Bucket metallurgy*

Left*: Plant tailings*

Middle*: Candelaria Mine adit*

Right*: The Candelaria Mine "El Tio"*

Underground

Left*: Pushing an ore cart*

Middle*: Dumping the contents of the cart*

Right*: Shoveling ore into bags to be hauled out*

Left*: A windlass hauls the bags up a level*

Middle*: A railway*

Right*: An apprentice guide demonstrates the use of explosives.*

View of 4,600 meter Cerro Chorolque as seen from the Altiplano; Alfredo Petrov photo, 1999

Tasna and Chorolque

by ***Jaroslav Hyršl*** and ***Alfredo Petrov***

Cerro (Mount) Tasna (or Tazna) and Cerro Chorolque are two prominent peaks protruding from the altiplano in southern Potosí department. Each rises above its surroundings and is easily visible from the other. They are 30 kilometers apart and display very different geologies: Tasna is composed mainly of tourmaline-bearing hornfels, whereas Cerro Chorolque is a volcanic neck, exposed by erosion. Nevertheless, both produce remarkably similar suites of ore minerals, each yielding tungsten, bismuth, tin and a little gold. Geochemically, both deposits exhibit well-defined temperature zoning, with bismuth and tungsten in a central high-temperature core, surrounded by a tin zone and finally by outlying lower temperature zones containing antimony, lead, and silver.

Cerro Tasna reaches an altitude of 5,100 meters (almost 17,000 feet) and hosts what is probably the largest bismuth deposit in the world; it is one of the most interesting mineral specimen localities in Bolivia and one of the most productive. Its bismuthinite crystals and ferberite twins are the world's best. The deposit also yields unusually fine specimens of pyrite, siderite, augelite, Japan-law quartz twins, native bismuth, and joséite-(A).

Collecting at this locality is not recommended for those who insist on their creature comforts; life on Tasna is harsh. There are no hotels or restaurants and no running water in the area; so visitors have no choice but to make do with whatever food is available in the market and are dependant on the kindness of miners for meals and lodging. Water is a precious commodity carried by schoolchildren from another mountain, about 2 hours away.

Cerro Chorolque is a 5,603-meter peak (well over 18,000 feet) where rich veins of cassiterite (and gold) are worked almost to the top of the mountain. It is, thus, one of the world's highest-altitude mines, employing an unusual mode of transportation: after entering the cage in an underground hoist room at the end of a long gallery, one is unexpectedly taken up towards the peak instead of down. Santa Barbara, one of the miners' villages on Chorolque, is situated at almost 4,600 meters (15,000 feet) and is perhaps the world's highest permanently inhabited community. Nighttime temperatures are below freezing throughout the year, and the houses have no heat. A miner there told the authors that he really enjoyed working night shifts in the winter, because while the temperature in the village sank to -20 degrees Celsius (-4 Fahrenheit), it remained a "balmy" 0 degrees Celsius (32 Fahrenheit) inside his adit! Water seeps into some of the galleries and freezes into ever thickening curtains of ice that have to be periodically blasted and laboriously mucked out. Although Cerro Chorolque has not produced as many world-class crystals as Cerro Tasna, it is at least as interesting mineralogically, yielding curiosities such as cassiterite epimorphs after various other species and micromounts of gold in unusual association with ferberite and scorodite.

Tasna Ore Minerals

Pinkish masses of native **bismuth** with perfect rhombohedral cleavages are common in superficial

parts of some bismuthinite veins. The specimens recovered in recent years have rarely exceeded 10 centimeters across, but they may contain inclusions of terminated quartz crystals, columnar bismuthinite, black ferberite, platy joséite-(A), pinkish fluorescent apatite, and rarely, small blebs of native gold.

Bismuthinite was, at one time, abundant, and very good crystals were found some decades ago; still it was a surprise when crystals reaching 10 centimeters in length were found in the winter of 2000. They are terminated by pyramidal faces showing strongly striated prism faces. Some crystals have holes in their terminations that, in some cases, pierce the entire crystal. In one pocket, the crystals were covered by a thin crust of marcasite that could be mechanically removed with some difficulty. A few crystals came out uncoated; these exhibit a strong metallic luster. Some of these crystals bear such a resemblance to the famous stibnite crystals from Japan that collectors are loath to believe they are bismuthinite (although bismuthinite can be easily distinguished by its much greater density). Some bismuthinite crystals are slightly weathered and are yellowish and dull.

In the winter of 2001, a new pocket yielded bismuthinite crystals up to 7 centimeters long that were completely covered by a black, feltlike coating of jamesonite, which can be mechanically removed. Specimens found at the end of 2002 have thin pyrite crusts and exhibit sharp terminations, but crystals reach lengths of only about 2 centimeters. Then in late 2008, miners recovered oxidized and corroded bismuthinite crystals over 30 centimeters long!

For a long time, **cassiterite** was the most important economic mineral on Tasna, but it only rarely forms euhedral crystals and those are only tiny druses. More commonly, Tasna's cassiterite is massive, although it sometimes occurs as heavy, knobby, waterworn, brown lumps, referred to as "stream tin," which are much appreciated by Bolivian collectors, who fondly call them *sapos* (toads). Bolivian dealers have offered perimorphs of cassiterite after apatite or bismuthinite from Tasna, but they more likely come from Cerro Chorolque.

Ferberite is presently Tasna's primary ore mineral, and the mountain produces very nice thick tabular crystals with a very bright luster. These are rich in crystal faces, fourteen of which have been described to date. Some of the crystals found in 2001 have unique parquet-like surfaces. Wonderful V-shaped twins are a Tasna specialty. The twins, which can reach about 15 centimeters in length, are generally found loose, with a very limited number of matrix specimens having been recovered.

Vugs lined with some of the most aesthetic ferberite specimens in the world were found in 2002. The platy crystals reach lengths of about 7 centimeters. Even some twins neared 4 centimeters in length. Quartz crystals to 5 centimeters, bismuthinite crystals coated by black jamesonite, pyrite balls, and (rarely) white apatite crystals were associated with the ferberite in some of the vugs.

Collectors of rare species have been thrilled by the platy, unusually large (up to 3 centimeters) **joséite-A** crystals embedded in native bismuth that came out in 2005 and 2006. Very nice **pyrite**, often

Above left: *An indigenous miner standing inside an ice-covered gallery of the Chorolque Mine; Gary Moss photo, 1999*

Above right: *A 9.2 cm wide bismuthinite crystal group from the Tasna Mine; Bill and Carol Smith collection; Jeff Scovil photo*

Above: Siderite from Tasna; 9.8 cm wide; Kosnar specimen; Brian Kosnar photo

Lower left: *Pyrite from Tasna; 6 cm wide; Rock Currier collection; Jeff Scovil photo*

Lower right: *Ferberite twin from Tasna; 6 cm wide; Bruce Carter collection; Jesse La Plante photo*

with unusual convex and diploidal faces, has been found in many veins. An April 2000 find produced very aesthetic and lustrous crystals up to 4 centimeters in association with siderite and augelite.

Tasna Gangue Minerals

Cerro Tasna is one of the Bolivian tungsten mines that produces rare Japan-law twinned **quartz**. These flattened crystals can be as much as 3 centimeters wide and sometimes form in association with V-shape ferberite twins. The locality has also produced some of the world's best **siderite**, in a wide range of habits. One spectacular specimen consists of a rhombohedron, more than 7 centimeters on edge, which is made up of many smaller crystals of the same habit. A 2000 find produced large druses of lens-shaped yellow-brown crystals up to 5 centimeters in combination with pyrite and augelite.

Augelite was first discovered on Tasna in April 2000. To date, it has always been found in association with pyrite and siderite. The white or green-tinged crystals can reach about 1 centimeter. Very rare gray pseudomorphs of **woodhouseite** after apatite were found on some bismuthinite druses in 2001. These thin, hollow casts after tabular or lenticular **apatite** crystals reached about 2 centimeters in diameter.

Tasna Secondary Minerals

Cerro Tasna is the type locality for **daubréeite**, which was described by Ignacio Domeyko (1802–1889) in 1876. It normally occurs mixed in with other secondary bismuth minerals, which are common in the superficial parts of many veins. X-ray analysis confirmed that daubréeite, bismutite, bismite, bismoclite, preisingerite, and similar-looking plumbojarosite are all found in Cerro Tasna veins, which occur at the top of the mountain. These secondary minerals are yellow to brown, powdery, and often form as pseudomorphs after bismuthinite. In some veins, they have been abundant enough to be exploited as ore. Canary yellow powdery **tungstite** is a common alteration product of ferberite, especially in the Farellon vein. It occasionally forms pseudomorphs after ferberite. A blue crust found by the authors in 1995 in the tailings of the Dorado Mine was identified as **aheylite**. It is one of only a handful of localities around the world from which aheylite has been confirmed to occur. A few rare

pale green pseudomorphs after bismuthinite were found to be composed mainly of **waylandite**. Greenish grey massive crusts of **scorodite** also occur.

Cerro Chorolque Minerals

Small blebs and tiny wires of **native gold** are quite common on Chorolque. These generally occur in cassiterite or sulfides. Round spots and films of supergene gold occur with **scorodite** and **jarosite** on ferberite in fissures in the oxidation zone. **Bismuthinite** is common in some veins but has been replaced, in superficial parts, by yellow-brown secondary minerals, mainly **bismutite**. **Cassiterite** is the most important mineral, and although it does not occur as nice crystals, it often forms very interesting paramorphs (casts) after older crystals of other minerals that have since disappeared. The original minerals, identified by the shape of cavity they left, were apatite, pyrite, and bismuthinite (Hyršl and Petrov, 1998a). These specimens can be quite large, weighing up to a few kilograms. Apatite- and pyrite-bearing cavities can reach about 2 centimeters; bismuthinite is occasionally quite a bit larger. **Apatite** crystals partly replaced by cassiterite have been very rarely found. **Siderite** is common and sometimes forms nice rhombohedra, which can reach 3.5 centimeters. A huge specimen, with sharp gray rhombohedra up to 5 centimeters, is on display at the British Museum of Natural History in London.

Large cave-like vugs on Chorolque contain stalactites of white **opal**, usually with limonite/jarosite stalactites. Similar vugs occur at two other Bolivian mines: Huanuni and Colavi. See the sidebar on Huanuni (page 23) for an explanation of how these unusual speleothems form.

***Top**: Ferberite from the Tasna Mine; 4.8 cm tall; Mike Bergmann specimen; Jeff Scovil photo*

***Middle**: Miner's wives and children offering specimens to Tasna visitors; Alfredo Petrov photo, 1999*

***Below left**: Alfredo Petrov with miners at the Chorolque Mine entrance; Gary Moss photo, 1999*

***Below**: Bismuth from Chorolque; 12.5 cm wide Arkenstone specimen; Joe Budd photo*

Cerro Huañaquino

Alfredo Petrov, whose appreciation for Bolivia's black minerals is legendary, was the first person to report magnetite crystals on Cerro Huañaquino.

***Upper**: Cerro Huañaquino; the white streaks on the steep slope to the right are the magnetite dumps; Alfredo Petrov photo*

***Lower left**: Magnetite from Cerro Huañaquino; 5.5 cm wide; Paul Stahl collection; Jeff Scovil photo*

***Lower right**: Apatite-(CaOH) from Cerro Huañaquino; 4.1 cm tall; Kosnar specimen; Brian Kosnar photo*

Magnetite and Apatite

Agustín Arando, a Potosí mine owner, prospector and mineral dealer, visited the author in 1993, bringing a small sack filled with some rather unusual loose apatite crystals, ranging from 2 to 8 centimeters long. Most have opaque gray to glassy yellow interiors and white, silky outer crusts, and they commonly have holes running down the centers—a few could even be blown as natural whistles! Many of the crystals serve as the matrix for abundant, tiny pseudomorphs of reticulated rutile after ilmenite. Some of the apatite crystals also play host to tiny smoky quartz crystals and rosettes of tabular barite crystals. These are all unusual associations and are certainly something new for Bolivia.

Agustín said that a campesino from the village of El Molino, about a half hour drive north of the city of Potosí, had brought him the crystals but that he refused to divulge their locality. I told him that there would likely be other interesting things wherever these were from; so he attempted to extract the locality information from the campesino. The man refused to drop a hint, but claimed that they were from a place far away, and that it required several days to hike there. Agustín tried diligently, discreetly following the fellow around and even hiring someone else to follow him, but with no success.

In my opinion, the several-days'-hike story was deliberate fiction: campesinos rarely hike onto the territory of other villages—it is a good way to get lynched—so I thought that the occurrence was probably not far from El Molino. Thinking that perhaps a phosphate-rich deposit would enable richer plant growth in that region of scrawny vegetation and chronic overgrazing, I studied Landsat images and aerial photographs looking for spots with thicker vegetation. Indeed, I found such a patch high up

Magnetite and Stromatolite

among the peaks of Cerro Huañaquino, a steep, craggy, eroded, light grey dacite intrusion that sticks out above the Cretaceous sedimentary red beds around it. Agustín hiked up, found a large apatite vein outcrop, and promptly registered a mining claim under the name "David Mosiah," a reference to his Mormon religion. A small pit of unknown age indicated previous exploratory work, probably by someone whose hopes of finding a tin or silver deposit were dashed.

One day in 1994 as I accompanied Agustín up to his claim (an exhausting 3-hour climb up to 4,400 meters), I noticed a loose, thumbnail-sized, octahedral magnetite crystal lying in the sand of a gully. This was the first of many magnetite finds on Cerro Huañaquino. Several tonnes of fine magnetite specimens have since been exported to specimen wholesalers in California, Arizona, and Germany. Many collectors say the crystals are reminiscent of the classic Swiss magnetite, although the matrix is totally different—the Bolivian matrix is volcanic. There is no iron ore deposit here.

On Cerro Huañaquino, magnetite crystals encrust the volcanic rock in narrow fissures that are mined exclusively for specimens. Adits penetrate a few meters into the dacite, but the dacite gets much tougher as the miners pass the weathered zone. Once the work gets too difficult, they give up and start a new adit elsewhere. The most prolific producer has been the San Calixto claim, named for former claim owner Calixto, a native of the village of Totoral, on the other side of the mountain. (NB: Some mineral labels erroneously list his home village as the magnetite locality.) Since 2008, the magnetite workings have been paralyzed by an ownership dispute between Agripina, Calixto's widow, and her neighbors.

Surprisingly, Agustín Arando's David Mosiah claim yields many microcrystals of magnetite among the big apatites, but no larger magnetites. Small apatite crystals occur among the larger magnetite specimens in several magnetite veins on other claims, but large magnetite crystals do not seem to occur in the same veins as large apatite crystals.

Pond Scum

The river valley below Cerro Huañaquino shows long parallel ridges of dredge tailings. The alluvium had been worked by a dredge that floated in its self-made pond, concentrating cassiterite sand that had washed down from Cerro Rico, about 20 kilometers

Egg carved from El Molino village 1stromatolite rough; 6 cm tall; Rock Currier collection; Jeff Scovil photo

upstream. The foothills of Cerro Huañaquino consist of multicolored Cretaceous marls; some strata are rich in clay and gypsum, others in limestone and fossil fish. One 40-centimeter-wide bed has produced a lapidary product distributed around the world: stromatolite (aka fossil pond scum), known in Bolivia as *pucalitos* (from *puca*, the Quechua word for "red" and the name of the geological formation hosting the red marls, and *lithos*, for "stone").

An Italian dealer at the Munich show once displayed a whole table of Bolivian pucalitos items under a banner that read "Precambrian stromatolite." When asked why he called them "Precambrian," the dealer dismissively stated, "Everybody knows stromatolites are Precambrian." While stromatolites are one of the oldest forms of life on the planet, one wonders what this dealer would think of the dinosaur tooth found embedded in a Bolivian stromatolite or the still-living stromatolites in Shark Bay, Australia.

In Bolivia, Pleistocene stromatolites can be found around the now dry shores of the Salar de Uyuni, and Cretaceous stromatolites are found in several parts of the Altiplano and Eastern Andes. But only two localities have produced lapidary-grade stromatolites: the village of El Molino in the foothills of Cerro Huañaquino and a hill in the Altiplano, 8 kilometers west of Sevaruyo in Oruro department.

***Background**: Cerro Huañaquino stromatolite slab (image screened); 29 cm tall; Fred Olsen collection; Jesse La Plante photo*

Llallagua!

Authors ***Jaroslav Hyršl*** *and* ***Alfredo Petrov*** *based this chapter on an article that they published in* The Mineralogical Record *in 2006.*

Llallagua is a tin deposit on Cerro Salvadora in Bustillo province, northern Potosí department in the eastern Andes. It sits at an altitude of more than 4,000 meters (13,000 feet). At its peak in the middle of the twentieth century, Llallagua was the best-known and productive tin mine in the world. The mine is still active, and specimens regularly reach the collector markets in North America and Europe.

Properly speaking, "Llallagua" (pronounced yah-YAH-wa) refers to the city in which the mine is located, but for this deposit, it is also the most generally used name on mineral locality labels. However, specimens from Llallagua sometimes appear in collections under a variety of loosely interchangeable locality names, including "Llallagua," "Siglo Veinte," "Siglo XX," "Uncía," or "Catavi." "Catavi" was the name of the nearby mill and was also the name of the administrative division of COMIBOL, which controlled the mine when it belonged to the State. *Siglo Veinte* and *Siglo XX*, both meaning "Twentieth Century" in Spanish, is the actual name of the mine and was derived from the mine's having opened at the beginning of the twentieth century. Uncía is the town adjacent to Catavi where the old smelter was. Other names that may appear on labels, especially in Bolivian collections, are "Salvadora," for the mountain which hosts the ore body, and the names of the individual miner-owned cooperatives (secciones) on different parts of the mountain, including "Cancañiri," "Dolores," "Dolores Atrás," etc.

To the consternation of foreign collectors, local dealers often write only the name of an adit or vein on their labels, omitting mine names; for example, an old monazite specimen in the San Diego Natural History Museum, labeled simply "San Jose," is from the San José vein in Llallagua, not the famous San José silver mine (page 26) in the city of Oruro .

History

Llallagua is much richer in tin and poorer in silver than similar subvolcanic deposits of Potosí and Oruro (Hyršl and Petrov, 1996). Silver was, nevertheless, produced at Llallagua in the early nineteenth century (Bandy, 1944), although it is not known what the silver ore minerals were and reports of native silver can no longer be substantiated. Only traces of discrete silver minerals (e.g., andorite) are now present; so the silver ore minerals may have been franckeite and stannite (both of which commonly bear a few tenths of a percent silver) and/or sphalerite containing microinclusions of tetrahedrite. In the late 1990s, a galena vein with minor pyrargyrite was worked in neighboring town of Uncía, on the southeast side of Cerro Salvadora; so it is possible that such veins were extant during the early nineteenth century and have since been obliterated.

***Left**: Vauxite and quartz from the Siglo XX Mine in Llallagua; 3.1 cm tall; Tom Hughes collection; Jeff Scovil photo*

***Right**: Paravauxite and childrenite on wavellite from Llallagua; the crystal is 2.3 cm wide; Joan Massagué collection (ex Jim and Dawn Minette, ex Mark Bandy); Jeff Scovil photo*

A Mineralogical Paradise

The modern history of Llallagua starts with prospecting for tin at the end of the nineteenth century. One of the most remarkable men in Bolivian history was Simon I. Patiño, a *mestizo* born into humble circumstances in 1868 in the village of Santibañez outside the city of Cochabamba. According to a widespread story, he was working as an assistant in a rural general store in Cochabamba. Around 1900 in return for a small personal loan he made to a miner, Mr. Patiño received the deed for a small tin prospect called Juan del Valle on the southeast side of the mountain in Llallagua. He moved to the property in 1903 and after spending some time digging with his wife, found a fabulously rich cassiterite vein. Patiño opened the first mine on the southeast flanks of the mountain that same year.

The Chilean Compañia Minera de Llallagua was, at that time, operating a mine on the northwest side of the mountain; still, by 1905, Patiño's company had become one of the most important in Bolivia. One by one, Patiño began buying the other properties in Llallagua, finally acquiring all of them in 1924, when he bought out the Chileans and formed Patiño Mines & Enterprises Consolidated. Patiño was well on the way to becoming one of the wealthiest people in the world.

Patiño also acquired other important Bolivian mines (the Huanuni tin mine and Kami tungsten mines, for example) as well as tin mines in Malaysia and even some tin refineries in Europe. Along with Moritz (Mauricio) Hochschild and Carlos Aramayo, Patiño became one of Bolivia's three tin barons. These men were, in some ways, more powerful than the State itself.

For the four decades prior to 1952, these three mining conglomerates accounted for approximately seventy percent of all Bolivian mineral exports by value. Under their leadership, tin mining reached its peak in 1929, when Bolivia produced 49,191 metric tons of the metal.

Patiño controlled his empire from Europe, where he had moved in 1912, and he seldom returned to Bolivia. By the end of the 1930s, Patiño's companies controlled more then 60 percent of the world's tin market. His descendants married European nobility and presently live in Switzerland and France. Simon Patiño died in Buenos Aires, Argentina, in 1947; his tin empire passed to his children, but only for the succeeding 5 years.

Many Bolivians resented the fact that the wealth generated by Bolivia's tin, laboriously torn from the ground by tens of thousands of underpaid miners whose lives seemingly had no value, was being squandered in Europe on fancy living. Rebellious sentiments were simmering.

The mining method of block-caving was introduced in 1950, and Siglo XX became the largest, most productive and modern tin mine in the world.

***Left**: Metavauxite, paravauxite, vauxite, and childrenite from Llallagua; 3.5 cm tall; Bill and Carol Smith collection; Jeff Scovil photo*

***Right**: Wavellite from the Siglo XX Mine in Llallagua; 5 cm tall; Marty Zinn collection; Jeff Scovil photo*

Vintage postcard of the Llallagua Mine (today the Siglo XX) from the collection of Günter Grundmann; the photo was taken ca. 1910 by Luis D. Gismondi, La Paz.

Two years later, however, the Patiño family lost all of their Bolivian mines in the revolution and nationalization of 1952 (although the overseas parts of their empire lasted much longer), and Siglo XX fell under the control of COMIBOL.

The first few years after the revolution were trying times for the mine engineers, laboratory workers, and other technical staff because the miners' unions enjoyed enormous political power despite their member's lack of education and technical expertise. Administrative and technical departments were forced to take miners onto their staffs (with reportedly disastrous effects on laboratory glassware). Strikes and violence were frequent, with many miners supporting the communist party. Llallaguans still tell stories about activist miners hiding underground for weeks in remote galleries when the police or army came looking for them; their wives smuggled food to them by way of forgotten passages and stopes.

In the 1960s another problem arose: the organized bands of hundreds of *jucos* (from the Quechua word for "nocturnal raptor"). Jucos were unemployed miners who created fictitious mining claims on other mountains but in reality were sneaking underground into Llallagua to high-grade cassiterite and sell it to Llallagua's COMIBOL-owned Catavi mill. The miners were soon joined in their juqueo by thousands of indigenous campesinos, creating a big headache for COMIBOL.

Cornelius Bloot, the Dutch manager of Catavi, knew what was going on, but he bought the juco's concentrates anyway. He did this partly out of compassion and partly pragmatism, as he knew that the jucos would simply sell the concentrates to another mill if he refused to buy them. COMIBOL ultimately also turned a blind eye to the juqueo, which provided concentrates to the mill more cheaply than COMIBOL could mine it themselves; after all, the company did not need to provide medical or other benefits to the poor jucos!

In the meantime, the world price of tin had been falling since the end of World War II, when the US military no longer needed canned rations. Bolivian politicians sucked up the profits from tin exports, leaving little for COMIBOL to put back into the company, and COMIBOL was awash in red ink. COMIBOL finally left Llallagua in the mid-1980s, handing the mines over to the local miners. By that time, more than 800 kilometers of galleries had been opened in the deposit!

At present, about 5,000 miners work at Llallagua, but conditions are far removed from the modernity of the 1950s and 60s. The mountain is owned by five cooperatives of mostly indigenous miners working with no capital and little technical support. Miners earn as little as 60 US dollars per month. The trains, hoists, and crushers no longer work. Miners carry ore out in knapsacks, crush it by hand with curved steel rockers, and concentrate it themselves in buddles. If Georgius Agricola were to visit Llallagua today, he would feel right at home!

It is notable that, although Siglo XX is primarily famous for producing tin, it has at various times also produced lesser quantities of silver, tungsten (mostly during World War I), and bismuth (mostly 1923-1925). The total amount of tin produced in Llallagua is currently unknown, but it had yielded about 500,000 metric tonnes by the early 1960s (Ahlfeld and Schneider-Scherbina, 1964).

Geology

Cerro Salvadora, the mountain which hosts the Llallagua tin deposit, is a Tertiary-age, funnel-shaped subvolcanic stock, like those of Potosí and Oruro. On the surface, it forms a 1.7-by-1.1 kilometer ellipse, which narrows with depth; still, at the 650-meter-level, the ellipse is 1,000 by 700 meters. The stock is made up of porphyry and porphyry breccia, having likely originated as an explosive vent. Llallagua's stock, like Potosí and Oruro, is heavily altered, and quartz, sericite, clay, tourmaline, pyrite, and other minerals have replaced almost all of the original feldspar in the volcanic rock. Tourmalinization, however, played a much bigger role in Llallagua than it did in Oruro or Potosí. In fact, Llallagua may represent the world's largest deposit of tourmaline, although mostly microfibrous and never gemmy.

Forty-seven main veins and approximately 1,500 rich, smaller veins are known in the deposit. Most of them trend northeast to southwest. Most of Llallagua's tin was produced from the veins, which transect the volcanic stock. Some of the veins (e.g., Contacto, Plata, Bismarck, and San Fermín), however, extend into the surrounding sedimentary rock, and the Carnevalito vein lies entirely in the sediments. The Contacto vein, perhaps the most important from a mineral collector's viewpoint, got its name from the fact that it crosses the contact between the igneous intrusion and sedimentary country rock.

Llallagua is a classic example of a "telescoped" deposit, meaning that high-temperature minerals were deposited close to the surface (i.e., low pressure). Low-temperature minerals were later superimposed on the same veins, resulting in the characteristic complexity of Llallagua mineralization.

The principal gallery, Siglo XX, on the edge of town, runs along level 650 of Cerro Salvadora; the richest mineralization is below this level. (Levels in Siglo XX are customarily designated by the number of meters below the peak of Cerro Salvadora.)

The earliest stage of vein mineralization is represented by quartz, tourmaline, bismuthinite, and cassiterite, with minor wolframite-group minerals (ferberite and hübnerite) and apatite. Some of the veins were once fabulously rich. Turneaure (1960) mentioned that the San José-San Fermín ore shoot, which had a maximum stope length of 700 meters and ore extending 600 meters vertically, contained 25 percent or more tin in many sections. The second mineralization stage deposited pyrrhotite and franckeite; and in the third stage, pyrite, marcasite, and siderite replaced almost all of the pyrrhotite.

Collectors covet the various phosphates found in the deposit. These phosphates, most abundantly wavellite, formed in the final stages of hydrothermal deposition and were derived from the decomposition of primary apatite. Greenockite and allophane were also deposited in this final mineralization phase.

Minerals

Two American mineralogists assembled the largest and best collections of Llallagua specimens: Dr. Samuel Gordon (in the 1920s) and Dr. Mark Bandy (in the 1940s). Bandy's collection now resides at the Los Angeles County Museum of Natural History in California. Anthony Kampf, the museum's curator, described the species jeanbandyite, which he found among the Llallagua specimens in Bandy's collection (Kampf, 1982).

More than 110 mineral species are known from Llallagua, but we will only cover those most interesting to collectors.

Allophane - $Al_2SiO_5 \cdot H_2O$

Pearly white mammillary masses of allophane form the matrix under some vauxite, vivianite and wavellite specimens. The allophane is unstable, and unless it is kept under water, it dehydrates to a chalky aggregate that gradually crumbles away, leaving vauxite specimens as thin crusts with empty mammillary impressions on the bottom.

Pictured is the back of a 12.5 cm wide vauxite crust (the front of the specimen is pictured on page 1) that formed on mammillary allophane. The unstable allophane dehydrated, leaving the white chalky material on the back of the crust.

Marty Zinn specimen (ex Marv Rausch); Jesse La Plante photo

Bismuth - Bi

Native bismuth was occasionally found in the deeper levels of the mines. It is younger than bismuthinite and the wolframite group minerals. The largest cleavable masses are 10 centimeters across. In the early 1990s, an interesting lot of specimens came from about level 450 of the San José vein, where small masses of native bismuth were associated with blue spherules of variscite.

Bismuthinite - Bi_2S_3

Bismuthinite is one of the oldest minerals in the veins and includes some of the largest known crystals of this species. It forms lathe-shaped crystals to 30 by 1 centimeters (Gordon, 1944) in vugs filled with iron sulfides or cassiterite. Freestanding terminated crystals up to 16 by 4 millimeters can still be found (rarely, however) on the dumps.

Cassiterite - SnO_2

Cassiterite is by far the most important ore mineral in Llallagua, often forming huge vuggy crystal-lined masses. Specimens do not reach the size and beauty of the better-known specimens from the Viloco Mine in La Paz department. Cassiterite from Llallagua is frequently coated with younger drusy wavellite. Rare younger simple prismatic crystals and colloidal "wood-tin" are reported by Bandy (1944). Most wood-tin attributions to Llallagua are erroneous; specimens are much more abundant on a volcanic plateau in the Macha area, about 80 kilometers southeast of Llallagua. On the vast dumps of volcanic waste in Llallagua, it is easy to find vugs with tiny, translucent, dark brown cassiterite crystals associated with quartz and tourmaline.

By the time Mark Chance Bandy (1900–1963) arrived in Bolivia in 1936 to work for the Patiño Family, he was already a seasoned mining engineer and mineralogist. He held a degree in mining engineering from Columbia University and was well on his way to obtaining a Ph.D. in mineralogy from Harvard (which he was awarded in 1938). His experience included five years working at Chuquicamata, Chile, from which he brought back several new mineral species, one of which, bandylite, was subsequently named for him.

The Patiño Family initially put Bandy to work exploring mineral deposits throughout Bolivia, which afforded him the opportunity to assemble a substantial and diverse collection of Bolivian minerals. Before long, he settled in as chief geologist, chief engineer, and finally general manager of the mine at Llallagua, where he focused his attention until leaving Bolivia in 1947.

Llallagua, at that time, was the world's largest tin mine and the source of many exotic and well-crystallized mineral species. It's hard to imagine a more exciting and rewarding position in which a mineralogist and professed lover of minerals could find himself. Bandy dedicatedly put his mineralogical acumen to work in documenting the many mineral occurrences encountered during the mining, and he endeavored to preserve as many finely-crystallized and unusual minerals as he could. In 1944, he published the results of his work in a monograph entitled *Mineralogy of Llallagua, Bolivia*.

The bulk of the mineral collection of Mark Chance Bandy was donated to the Natural History Museum of Los Angeles County by his widow Jean in 1977, with an additional portion coming in 1980, and the remainder being transferred after Jean's death in 1991. The collection includes more than 700 specimens from Bolivia representing 40 localities, but by far the largest portion (about 80 percent) is from Llallagua.

—Anthony R. Kampf

Chalcanthite - $CuSO_4 \cdot 5H_2O$

Masses of beautiful blue crystals of post-mining chalcanthite occur in several sections of the mine, especially along the San José vein. Chalcanthite is found in fissures in the rock walls or hanging from wooden beams. The latter are popular with local mineral collectors, although many of their specimens of *calcantita* are really the more common cuprian melanterite (pisanite variety).

Childrenite - $Fe^{2+}Al(PO_4)(OH)_2 \cdot H_2O$

Childrenite used to be found as druses of millimeter-sized tabular brown crystals coating extensive fissure surfaces. It has recently been found as very dark brown, radiating spheres reaching 3 millimeters in diameter. These commonly have whitish centers and sometimes occur on vivianite. Local Bolivian dealers have erroneously sold childrenite as sigloite, a misidentification perpetuated by some foreign dealers; caveat emptor!

***Right**: Childrenite on quartz from the Siglo XX Mine; 3.3 cm wide; Kosnar collection; Brian Kosnar photo*

Cronstedtite - $Fe^{2+}_2Fe^{3+}(Si,Fe^{3+})O_5(OH)_4$

Black crystals of cronstedtite, with triangular cross sections and excellent luster, are occasionally found on hisingerite, marcasite, pyrite, siderite, or rhodochrosite. According to Bandy (1944), the best crystals came from the 411 and 516 levels of the Contacto vein. The species is abundant on recent vivianite specimens as olive-green mammillary crusts associated with pyrite and sphalerite.

Cylindrite - $FePb_3Sn_4Sb_2S_{14}$

Cylindrite is more famous from Poopó, its type locality, but Llallagua produced unique terminated rods up to 1 centimeter long in vugs that are always associated with franckeite. Crystals are still occasionally found in recently produced vuggy franckeite ore. The best-known cylindrite specimen from Llallagua is on stannite and is in the Los Angeles County Museum of Natural History.

Diadochite - $Fe_2^{3+}(PO_4)(SO_4)(OH) \cdot 6H_2O$

Vitreous stalactitic masses of diadochite, of colloidal origin, were found on level 320 of the Contacto vein, and large masses have been found in other stoped tin veins. Bandy (1944) reported that coffee brown diadochite had formed in great abundance on the floors of old workings below the 650 level. In fact, the formation was up to 60 centimeters deep in parts of the 720 level, which had been flooded for over ten years. It was sometimes plastic when first found, but it dehydrated rapidly outside the mine and crumbled into pieces smaller than 1 centimeter.

Evansite - $Al_3(PO_4)(OH)_6 \cdot nH_2O$

Light orange, transparent, resinous masses of evansite, which resembles tree resin, are occasionally found in association with metavauxite.

Potassium Feldspars - $KAlSi_3O_8$

Fresh unaltered orthoclase and plagioclase (probably "oligoclase"), frequently as Carlsbad-law twins, have been found only sporadically as phenocrysts in the volcanic porphyry. Almost all of the Carlsbad-twinned K-feldspar crystals have been replaced by younger minerals, usually tourmaline, sericite, quartz, kaolinite and, rarely, cassiterite, or pyrite (Hyršl and Petrov, 1998). These pseudomorphs can reach up to about 8 centimeters.

Florencite-(Ce) - $CeAl_3(PO_4)_2(OH,H_2O)_6$

Creamy white florencite crystals up to 1 centimeter wide were found a few years ago in association with pyrite and color-changing monazite-(Ce) at the Sección Dolores gallery.

Fluorapatite (Apatite-(CaF)) - $Ca_5(PO_4)_3F$

Fluorapatite is one of the most beautiful Llallagua minerals and used to be quite abundant in colorless, purple, or pink tabular crystals, although fine specimens are almost unobtainable today. Some small crystals are gemmy. Metallic-looking crystals, heavily included with fine jamesonite needles, have also been found. The largest transparent to translucent, colorless crystals, which were collected by Bandy in the Contacto vein in the 1940s, reached 5-by-2 centimeters and had nine crystal forms! Bandy spent several days carefully collecting apatite crystals from a very large vug, but some of his fellow mine engineers resented the amount of time it took away from his regular work. One of his companions got so angry that he destroyed the vug with a blast of dynamite after Bandy had only collected half of it.

Llallagua apatite fluoresces strongly yellow-orange on the {1010} faces and bright violet on the {0001} faces. This was studied in detail by John Rakovan (2003) who also determined the mineralizations to have occurred 43.8 +/- 4.7 million years ago. The latest find, in 2002, yielded tabular apatite crystals to 3 centimeters wide but only about

***Top**: Fluorapatite from Llallagua; 2.7 cm wide; Les Wagner Jr. collection; Jeff Scovil photo*

***Above left**: Cronstedtite from the Siglo XX Mine; 4.6 cm wide; Kosnar collection; Brian Kosnar photo*

***Above right**: Fluorapatite from the Contacto vein in Llallagua; 4.7 cm wide; Natural History Museum of Los Angeles County collection (ex Mark Bandy); Anthony R. Kampf photo*

4 millimeters thick that are completely encrusted with yellow wavellite. They are light pink in daylight and tungsten light but lemon yellow under fluorescent tubes, a phenomenon that indicates high contents of certain rare earth elements.

Fluorapatite is one of the oldest minerals in the tin veins, and by leaching and decomposition, it gave rise to the numerous younger phosphates, especially wavellite and the vauxites group minerals. Alteration and replacement of the apatite gave rise to a variety of pseudomorphs: the wavellite pseudomorphs after apatite can be quite spectacular, and the authors have also found apatites partially replaced by blue vivianite or white gypsum.

Franckeite - $Pb_6Sb_2FeSn_2S_{14}$

Franckeite commonly forms tabular to platy crystals, usually in association with marcasite and wurtzite. Vuggy franckeite masses, which are lined with drusy franckeite crystals ranging from 2 to 5 millimeters, are presently the main ore mineral in one section of the Contacto vein. Here it is associated with pyrite, arsenopyrite, wurtzite as well as rare cylindrite and hinsdalite.

Greenockite - CdS

Greenockite is widespread in small quantities as tiny pyramidal crystals, which are sometimes tightly grouped as thin crusts. It rarely occurs as rare cyclic

Above: *Greenockite and quartz; 5.5 cm tall; Natural History Museum of Los Angeles County collection (ex Mark Bandy); Anthony R. Kampf photo*

Right: *Monazite exhibits dramatic color change from grey-green under fluorescent light to beige under daylight to pink under incandescent light. This plate of monazite twins with quartz and cassiterite is 9 cm wide and is in the collections of Natural History Museum of Los Angeles County (ex Mark Bandy). Anthony R. Kampf photo*

quadruple twins—the classic Llallagua tetrapods. As in other Bolivian deposits, greenockite from Llallagua is bright brick red, in contrast to the yellow in which it is found in other worldwide occurrences. Greenockite appears to be a late-stage primary hydrothermal mineral and is not a decomposition product of cadmium sphalerite. Specimens with minute crystals sprinkled on wavellite, quartz, cassiterite, or marcasite are widely distributed throughout micromount collections.

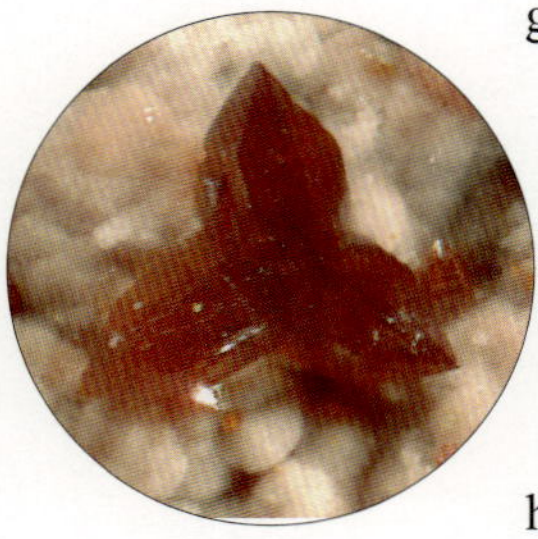

Halotrichite - $Fe^{2+}Al_2(SO_4)_4 \cdot 22H_2O$

Both halotrichite and pickeringite are postmining formations in old galleries. Mark Bandy observed that they started developing as loose fibrous masses on walls in dry temperate sections of the mine almost immediately after workings were abandoned (Bandy, 1944). Stalactites can reach lengths of 2.5 centimeters after 3 weeks and 20 centimeters in a year! Some miners erroneously tell visitors that this hairy stuff is asbestos.

Jeanbandyite - $(Fe^{3+}_{1-x} \square_x)(Sn_{1-y} \square_y)(OH)_6$ where $(3x+4y=1)$

Jeanbandyite was described by Anthony Kampf in 1982 and named after Jean Bandy, who had suffered several years in the high, cold, and dirty mining camp of Llallagua while her husband Mark worked there. After his death, Jean donated his collection to the Los Angeles County Museum, where Kampf discovered the new species on several specimens of Bandy's apatite crystals on stannite.

Jeanbandyite is orange-brown and forms small (approx. 0.2 millimeter) intergrowths with yellow octahedrons of wickmanite. This is probably the same material that Bandy had prematurely named "patiñoita" (Bandy, 1976) after his boss. Jeanbandyite has since been found by the authors on the dumps in a completely different paragenesis: as crude orange octahedra covered by an orange plumbogummite crust, associated with hemispheres of banded "wood tin" cassiterite, quartz, pyrite, and muscovite.

Marcasite - FeS_2

The most interesting Llallagua marcasite specimens are tabular pseudomorphs after pyrrhotite up to 20 centimeters wide that are covered with wavellite. It has also been found as paper-thin crystals interleaved with similarly thin franckeite. In Bandy's days when deeper sulfide-rich levels were being worked, marcasite was abundant. Its decomposition has caused serious problems with the specimens in the Los Angeles County Museum. Marcasite is fortunately much less common on the upper levels of the deposit, which are currently producing specimens.

Left: *Metavauxite on Vauxite from the Siglo XX Mine in Lallagua; 2 cm tall; Arkenstone specimen; Joe Budd photo*

Inset: *Greenockite "tetrapod" twin; field of view 0.7 mm across; Natural History Museum of Los Angeles County collection; Anthony R. Kampf photo*

Melanterite - $FeSO_4 \cdot 7H_2O$

Beautiful green melanterite stalactites are common throughout the mine. The mineral also occurs rarely as crystals reaching 2 centimeters. The deep blue cuprian variety pisanite, another common post-mining product, is often mistakenly labeled "chalcanthite" by the miners.

Metavauxite - $Fe^{2+}Al_2(PO_4)_2(OH)_2 \cdot 8H_2O$

Metavauxite, described by Gordon (1944), is by far the rarest of the "vauxites." It forms white to light green acicular crystals, sometimes in radial aggregates. A few very nice specimens with needles up to 3 centimeters were found in summer 2001; these are among the best known for this species.

Monazite-(Ce) - $(Ce,La,Nd)PO_4$

Tiny monazite crystals were found widespread in small quantities in all of the veins. The largest reported crystals were embedded in iron sulfides in the Bismarck vein; they reach 2.5 centimeters (!) in length (Bandy, 1944). Llallagua monazite often appears pale pink in daylight and deeper pink under tungsten bulbs, but greenish grey to white under fluorescent tubes. This effect is caused by the significant quantities of neodymium that they contain. Unlike other occurrences, Llallagua monazite is practically thorium-free and is, therefore, hardly radioactive at all. Samuel Gordon reported thorium contents of less than 10 ppm (Gordon, 1939). Nevertheless, this trace is still sufficient to

Left: Paravauxite; 3.1 cm wide; Collector's Edge, Inc. specimen (ex Philadelphia Academy); Jeff Scovil photo

Right: Roberto "Loco" Alarcón, this Llallagua miner, died in the mine only two months after guiding the photographer to his work area. Photo Jaroslav Hyršl, 1999

Above: Vauxite on Paravauxite from Llallagua, 5.3 cm wide; Marty Zinn collection; Jeff Scovil photo

produce an autoradiograph after three months of exposure. Abundant grains and crude monazite crystals to 1 centimeter associated with white florencite crystals, quartz and pyrite were found in 1999 in the Dolores gallery. Lovely twins of this color-changing monazite can still be found on the dumps and a cavity with few dozens of specimens with excellent monazite twins, reportedly reaching 1 centimeter, on quartz crystals was found in 2008.

Paravauxite - $Fe^{2+}Al_2(PO_4)_2(OH)_2 \cdot 8H_2O$

Gordon described both paravauxite and vauxite in 1923. Paravauxite is the most common of the three vauxites group minerals and sometimes forms as large masses and crystals in vugs. The characteristic triclinic crystals are normally pale green when larger and colorless when forming smaller crystals. Surface oxidation can turn them pale yellow, but white opaque crystals are pseudomorphs of sigloite after paravauxite. Paravauxite has been fairly abundant in several recent finds, some yielding splendid light green crystals up to 3 centimeters.

Pyrite - FeS_2

Pyrite is very common and usually forms beautiful, sharp, octahedral crystals or sometimes cubes or combinations of cubic and octahedral faces—but never pyritohedra. Pyrite also appears as pseudomorphs after pyrrhotite and rarely after teallite or feldspar. Crystals from some weathered dumps exhibit peculiar tunnels or serpentine "worm holes" etched into them. The finest specimens were found around the periphery of the main mineralization zone. Nice pyrite crystals up to 7 centimeters were discovered recently in the Dolores gallery. These floaters are found embedded in soft, massive franckeite.

Pyrrhotite - $Fe_{1-x}S$ (x=0.1–0.2)

Pyrrhotite was a widespread primary iron sulfide; however, almost all was replaced by pyrite and marcasite. These pseudomorphs can reach an impressive 20 centimeters across and may be encrusted with wavellite. Bandy (1944) mentioned a portion of the Contacto vein on level 516, where unaltered pyrrhotite formed masses 2 meters wide.

Quartz - SiO_2

Quartz is the most common Llallagua mineral, with crystals reaching 10 centimeters in length. It can commonly be distinguished from quartz from other localities thanks to the crust of wavellite that commonly covers it. Flattened Japan-law twins occur in the centers of veins. Transparent quartz crystals with tiny metallic inclusions are common, the inclusions are often arsenopyrite or pyrrhotite and occasionally wolframites. Bandy (1944) described cassiterite and tourmaline embedded in quartz. Yellow-orange inclusions of monazite or xenotime up to about 2 millimeters wide are rare.

Rhodochrosite - $Mn^{2+}CO_3$

Llallagua is the only Bolivian locality that yields rhodochrosite. Nice, pink rhombohedra up to 1 centimeter form in vugs with wurtzite and sphalerite, but it has never been common. Some Llallagua rhodochrosite is pale brown rather than pink and is, therefore, difficult to distinguish from siderite.

Siderite - $Fe^{2+}CO_3$

As in most Bolivian mines, siderite is by far the most abundant carbonate gangue mineral in Llallagua, especially in younger veins and on the periphery of the deposit; still, crystals are much less common here than in other Bolivian tin mines. Siderite is found in association with iron sulfides, sphalerite, franckeite, and wurtzite. Nice rhombohedral crystals are rare.

Sigloite - $Fe^{3+}Al_2(PO_4)_2(OH)_3 \cdot 7H_2O$

Sigloite was described by Hurlbut and Honea (1962) as white to yellow, opaque pseudomorphs after paravauxite. Sometimes it occurs on orange childrenite. Sigloite is probably the same material that Bandy referred to as "hydrated paravauxite" (Bandy, 1976). Some misidentified "sigloite" on the species market is actually spheroidal brown childrenite. Crusts of light brown sigloite crystals are actually quite common but were discarded in the past as "weathered paravauxite."

Sphalerite - ZnS

Sphalerite is common, sometimes in crystals to 1 centimeter, and is generally wholly or partially replaced by stannite. Abundant small crystals of brown-black "marmatite" sphalerite are occasionally found on the faces of larger pyrite crystals. Herzenberg (1933) described raspberry red nodular or reniform masses of an arsenic-bearing sphalerite (0.64 weight percent As) of colloidal origin. He referred to the variety as "gumucionite," after Julio Gumucio, who was chief mining engineer at Llallagua and the only native-born Bolivian to assemble an important mineral collection. The red-pink color may be due to admixed realgar, although this is not certain. Pale pink to light orange porcelaneous sphalerite masses form parts of the matrices of some recent vivianite specimens.

Stannite - Cu_2FeSnS_4

Stannite from Llallagua is much less common than it is from Oruro, but Llallagua's few surviving specimens are very probably the world's best, even better than the new ones from Yaogangxian, China. Bandy (1944) described crystals reaching 4 centimeters with sphalerite, franckeite, and terminated rods of cylindrite. The best specimen is perhaps the one preserved at Harvard University. An excellent specimen from Bandy's collection is exhibited at the Los Angeles Museum of Natural History. Sharp microcrystals can be found on some franckeite specimens from the 1990s.

***Upper left**: Quartz twin; 3.4 cm wide; Tom Hughes collection; Jeff Scovil photo*

***Upper right**: Vauxite on quartz; field of view 6 cm wide; Paul Stahl collection; Jeff Scovil photo*

***Lower right**: Stannite from Llallagua; 6 cm wide; Natural History Museum of Los Angeles County collection (ex Mark Bandy); Jaroslav Hyršl photo*

Top: Vauxite; 22.5 cm wide; Carnegie Museum of Natural History specimen; Harold and Erica Van Pelt photo

Tourmaline Group Minerals

Tourmaline-group minerals are among the most common minerals in Llallagua; practically all of the original feldspars in the porphyry are replaced by fine-grained tourmaline. Sericitization of the feldspars was followed by tourmalinization. Bandy (1944) reported tourmaline constituting up to 90 percent of the black "breccia" dikes in the stock. Tourmaline also formed in the metamorphosed sediments around the stock. This process made the Llallagua intrusion one of the world's largest accumulations of tourmaline, although most of the tourmaline occurs as felty or fibrous microcrystal masses; none of it is in the form of cuttable crystals.

Some collector appreciate the tourmaline pseudomorphs after Carlsbad twins, up to 8 centimeters wide, beautiful. These can still be found on dumps and on the top of Cerro Salvadora. Acicular crystals, in several colors ranging from colorless to pale blue to light green to green-yellow and light brown to brown-gray are found on the dumps in vugs of altered volcanic stock. These are sometimes impaling spheres of younger phosphates.

Vauxite - $Fe2{+}Al2(PO_4)_2(OH)_2 \cdot 6H_2O$

Vauxite was first described by Samuel Gordon (1923) and was named for his sponsors, the Vaux family of Philadelphia. Blue crusts formed of tiny crystals are relatively common and are found in almost all of the veins. Drusy vauxite commonly forms on mammillary allophane, which crumbles to powder on dehydration, leaving a thin crust of pure vauxite with rounded impressions on the back. Very nice blue druses were found again in the summer of 2001 and were accompanied by pearly white allophane and pink crandallite. Vauxite spheres are much rarer but are sporadically found on quartz and cassiterite. Attractive green-blue to bright blue spheres of vauxite reaching 3 centimeters in diameter were found in 2003. Surprisingly, the authors found, by X-ray diffraction analysis, that yellow-brown to green-brown crystals from this new find are also vauxite; apparently oxidation is enough to change the color from blue to green-brown.

Vivianite - $Fe_3^{2+}(PO_4)_2 . 8H_2O$

Before the famous finds of vivianite crystals at Morococala and Huanuni in the 1980s and 90s, Llallagua was the source of the world's best specimens. The largest doubly-terminated crystals were found in clay and reached 7 by 2 by 2 centimeters. Surprisingly, nice specimens were found again in 2000; they are transparent, triangular tabular crystals up to 13 centimeters across, often occurring in parallel rows like saw teeth and associated with childrenite, cronstedtite, pyrrhotite, franckeite, and pink massive sphalerite.

Wavellite - $Al_3(PO_4)_2(OH,F)_3 \cdot 5H_2O$

Wavellite is one of the most abundant Llallagua minerals, usually found as white or pale yellow

coatings or crusts on quartz and cassiterite druses. Colorless single crystals up to 4 millimeters are common in the deeper parts of the mine. Quartz is sometimes covered by radial wavellite spheres, the largest reaching 4 centimeters in diameter. Wavellite also forms pale orange epimorphs after large tabular apatite crystals. Nice yellow wavellite spheres, to 1 centimeter in diameter, were found with apatite-(CaF) crystals on quartz in 2002. Llallagua wavellite can be colorless, white, pale yellow, or (rarely) orange, but it is never green like Arkansas specimens.

Wolframite Group Minerals

Wolframites were once an important Llallagua ore mineral. Minerals in the group were produced commercially as tungsten ore during World War I, but good crystals are much less common than they are from other Bolivian mines such as Tasna (page 44). Nevertheless, "fish-tail" twins are mentioned by Bandy (1944).

Wurtzite - $(Zn,Fe)S$

Well-formed wurtzite crystals are famous from several Bolivian deposits, and they were not uncommon in the secondary veins at Llallagua, which produced several habits, including fibrous and tabular crystals. The largest are pyramids up to 2 centimeters. A specimen composed entirely of black wurtzite crystals up to 10 millimeters, weighing 0.6 kilograms, is in a private German collection. A few tapering pagoda-like crystals, up to 2 centimeters tall, were found in vugs in pyrite in Sección Dolores in the mid-1990s.

Xenotime-(Y) - YPO_4

Dark yellow, prismatic xenotime-(Y) is much rarer than the associated monazite. Rarely, both form inclusions in quartz. Ahlfeld and Muñoz Reyes (1951) described crystals reaching 2 millimeters, but a fantastic specimen is exhibited in the British Museum of Natural History in London: a 7-by-4-centimeter matrix covered with yellow monazite crystals to about 9 millimeters and some smaller pale green xenotime-(Y) crystals. Transparent light yellow xenotime-(Y) occurs on a few of the recent pink monazite specimens.

Collecting Possibilities

Llallagua is a *cooperativa*, meaning that the miners jointly own it. Many thousands of very poor miners, laboring under very primitive conditions, are part of the cooperative. They exploit the remains of

Above right: *Wurtzite and quartz from Siglo XX Mine; 8.6 cm tall; Colorado School of Mines specimen; Jeff Scovil photo*

Upper left: *Wavellite on quartz; 6.7 cm wide; Les Wagner Jr. collection (ex Bandy); Jeff Scovil photo*

Lower left: *Entrance to the Siglo XX Mine; Jaroslav Hyršl photo, 1995*

View of Llallagua from from Cerro Salvadora
Jaroslav Hyršl photo, 1999

tiny underground cassiterite veins or are reworking old dumps. Only a few of them have any idea as to what a collectable mineral specimen should look like; thus, some miners decorate their homes with broken druses of quartz with cassiterite, but really well preserved specimens are uncommon.

The authors first visited Llallagua on a Saturday in 1993 and asked several miners for mineral specimens. The answer was always the same, "You are lucky, today is concentration day." We didn't understand until people brought us sacks full of powdered cassiterite then whined "Why don't you want it; it's so pure!" We were not able to get a single specimen during this first visit. Since then, a handful of miners have learned about the specimen market, and small lots of crystallized specimens, mostly phosphates, occasionally appear at mineral shows.

The good news is that collectors can easily collect on the enormous dumps, especially the mile or so of dumps stretching along the highway as one drives into Llallagua. The authors have found nice pseudomorphs of tourmaline after feldspar and pyrite after feldspar as well as vugs containing crystals of monazite, wavellite, florencite, cassiterite, jeanbandyite, variscite, tourmaline, bismuthinite, and other species on the dumps. Other interesting species are certainly still waiting to be field collected, but these will delight mainly micromounters, as the chances of digging a nice cabinet specimen out of the dumps are close to zero.

Incidentally, the mile of dumps along the highway, containing several million tons of material, are the remains of the "block-caving" procedure carried out from the 1950s to 70s. Block caving is the cheapest and most efficient form of underground mining, and Llallagua was one of the pioneers of this method, in which an adit is dug into the center of an ore body, then a large area of the ore body is undercut so that the roof begins to collapse. As the roof collapses and crashes to the floor, the rock breaks into small pieces, saving a lot of expensive trituration. The broken rock is mucked out through the access adit, and more of the roof caves in as the loose material from the floor is hauled away. As one can imagine, this is extremely dangerous work, and the death rate among miners was high—not that anyone cared much in those days. The fracturing of the roof can continue up to the surface, leaving a collapse feature resembling a volcano crater. In fact, there are several large craters of this nature, with frightening vertical walls, near the peak of Cerro Salvadora.

The rubble taken out of the mountain undergoes a sink-and-float process to remove the purer cassiterite and the rest of the rock is dumped, although it still contains a few tenths of a percent of cassiterite. The block caving dumps in Llallagua are remarkably clean, and collectors can lie on them without getting their clothes dirty. With an inexhaustible quantity of material to search through, the dumps are a collectors' paradise!

For a modest remuneration and with permission from whichever Cooperativa office is responsible for the sección one wishes to enter, Spanish-speaking collectors can arrange to accompany a miner underground to a working stope for some underground collecting. Be forewarned that Siglo XX is dangerous underground. Roberto "Loco" Alarcón, a friend of the authors, died during a bumpy ride to the hospital after having broken multiple bones falling down a rotten ladder. You will get dirty, you may encounter extreme temperatures—both hot and cold. Safety rules will not be observed, and you will not be able to hold anyone responsible if you get injured. In addition, you might have to stay underground for your companion miner's entire shift, as you may be unable to find your own way back to the surface.

Colquechaca: "The Silver Bridge"

*By **Terry Wallace,** who was curator of the University of Arizona Mineral Museum and has published widely on minerals, particularly silver minerals. Terry spent a decade doing geophysical field work in Bolivia, Chile, and Argentina.*

A rare vintage postcard depicting Colquechaca at around the turn of the century, printed in 1905; Günter Grundmann collection

Some of the world's richest silver and tin deposits sit at elevations reaching 5,800 meters in Bolivia's eastern Cordillera. These include Oruro, Llallagua, Chorolque, Cerro Rico (Potosí), and Porco along with a relatively modest district called Colquechaca. The indigenous peoples of the region are the Quechua. In the Quechua language, *Colquechaca* means "silver bridge."

For a brief period between 1885 and 1920, the mines of Colquechaca produced spectacular specimens of pyrargyrite, stephanite, and argentite, along with the world's finest canfieldite (silver tin sulfide). Canfieldite is named after Frederick A. Canfield, an American mining engineer and mineral collector who visited Colquechaca in 1885 and 1887 and helped preserve the wonderful mineral specimens from the district.

History

Colquechaca is about 100 kilometers southeast of Llallagua, in extremely rugged terrain. The town is situated at an altitude of approximately 4,300 meters, and the surrounding mines are at elevations as high as 4,900 meters. The earliest written records mention the driving of a long tunnel in the San Bartolome Mine before the year 1700 (Peele, 1894). The tunnel was more than 450 meters long, and the historic mine dumps, called *desmontes* (waste rock) suggest that the area had been mined for decades. The San Bartolome Mine was operating in the middle of the nineteenth century, and two rich lodes were discovered nearly a kilometer from the tunnel entrance. Between 1862 and 1890, the mine was in bonanza (Peele, 1894), and the population of Colquechaca grew to more than 7,000.

Compania Colquechaca-Aullagas de Bolivia was incorporated in the mid-1880s, and was the prime mine operator in Colquechaca for 20 years. Although the mines were productive during this time, the mining methods were crude. Both the Colquechaca and Aullagas mines were extensive, and the ore zones far removed from the surface. Thus, the ores were typically sorted underground; rich material was bagged, brought to the surface, and sized with sieves. The richest ore (silver concentrations greater than 500 ounce per ton) was sent to Oruro for smelting. The poorer grade ores were sent to a local amalgamation plant, the Ayoma Reduction Works, located about 35 kilometers from Colquechaca. The ruins and waste spoils from Ayoma are an impressive sight today and are a testament to the huge quantity of rock that has been removed from the deep mines.

The silver crisis of the mid-1890s hit Colquechaca hard. Even though rich ore remained,

A rare Colquechaca stock certificate from 1886

Terry Wallace collection

its remote location and the demand for power to keep the deep mines dewatered shut down most of the district by the start of the twentieth century. In 1914, tin baron Simon Patiño bought several flooded silver-tin mines in the Colquechaca district. For a decade, Colquechaca flourished as a tin mining camp with significant silver production. By the 1930s, the mines had become depleted although mining continued until the mid-1980s, when the global price of tin collapsed.

Geology

The geology of the eastern Cordillera is the product of hundreds of millions of years of Earth's history. The Pacific coast of South America defines the western margin of the South American plate, which has been colliding with the oceanic Nazca plate to the west since Mesozoic times (more than 200 million years ago). Ocean-continent convergence results in subduction; one of the most profound consequences of the subduction of the Nazca plate is that as the slab descends, hydrated minerals (e.g., amphibole) and the water trapped in the oceanic crust are introduced to increasing temperatures and pressures. At some point, the hydrated minerals become unstable, and the water is driven out of them. The released water rises into the hot mantle overlying the Nazca slab, promotes melting, and creates massive volumes of magma. These magmas concentrate metals, and the magmatic fluids travel into the surrounding wall rock, filling veins and creating lodes that ultimately become the grist of great mines.

The detailed geologic evolution of the eastern Cordillera is complex, but from the standpoint of the silver and tin deposits, a huge outpouring of Miocene ignimbrites deposited acidic subvolcanics in the host Paleozoic clastic metasedimentary rocks. The intrusives have associated polymetallic veins and carry significant amounts of silver, tin, and zinc (Sillitoe et al., 1975). The rich veins in many of the great mines strongly suggest that the concentration of metals is not related to weathering but rather to the hydrothermal processes. Very little work has been done on the petrogenesis of the veins at Colquechaca, but Cerro Rico has been studied in detail. The veins at Cerro Rico are highly fractionated and enriched in incompatible elements such as silver and tin, suggesting that the mineralization period was prolonged, requiring long-term magma-related hydrothermal activity. There were probably multiple magma injections into the igneous complex from a high-level crustal magma reservoir.

The Mines of Colquechaca

Minerals labeled as being from Colquechaca commonly have a confusing litany of locality information that mixes veins, tunnels, mines, and towns. This is due partly to the remoteness of the district and partly to the manner in which specimens were collected and dispersed. An example of this is associated with the type description of canfieldite: Penfield discussed obtaining a specimen of "argyrodite" from William E. Hidden of New York that was labeled as being from La Paz, Bolivia. The specimen turned out to be canfieldite and must have been from Colquechaca, but because the mine owner lived in La Paz, the specimen was labeled as being from that city.

In 1894, Peele described five properties in the district and two principal mines: The Aullagas Mine and the Colquechaca Mine. The mines exploit a series of very steeply dipping veins. The most important of these veins are the Embudo, Gallofa, Empresa, Carmen, San Matias, San Augustin, and San Miguel. The veins are between 5 and 25 centimeters thick and contain vugs reaching about one meter across that carry the richest ore. The vein material is commonly banded, and the vugs are generally lined with crystallized quartz. The quartz often breaks easily from the country rock, making for the spectacular mineral specimens. Peele describes two classes of ore: *broza-guia*, "first class" and *broza*, "second class." The first class ore consists mainly of pyrargyrite, native silver, acanthite, and rarely, tetrahedrite. The ore is commonly associated with sphalerite and minor galena. The greatest silver value is carried by massive and crystalline pyrargyrite; "perfect" crystals to 10 centimeters are described as common. Typically, the first class ore runs from 500 to 5000 ounces of silver per ton. The second-class ores have similar mineralogy but carry less than 100 ounces of silver per ton.

The Minerals

Today, the town of Colquechaca is largely deserted. Over the road is an arch with a rusty sign that simply states "Ruby Silver Capital of the World." Although it seems a hollow boast, pyrargyrite, and to a lesser extent proustite, from Colquechaca are, in fact, among the finest ever found. Every great mineral museum has a fine Colquechaca pyrargyrite cluster, often on bright white quartz—literally thousands of fine specimens came out between 1885 and 1920. Although Colquechaca has not produced a great specimen for a century, many fine examples are still in circulation.

Much of what is known about the mineralogy of Colquechaca is the result of the work of two men: Friedrich Ahlfeld and Frederick Canfield. Ahlfeld, who is known as the "Father of Bolivian Geology," wrote extensively on the Bolivian tin, tungsten, and silver deposits, publishing 142 papers on Bolivia between 1925 and 1975. His most famous work, *Mineral Deposits of Bolivia*, was first published in 1939. Ahlfeld built several important mineral collections. An early collection was purchased in 1932 jointly by Harvard University (Cambridge, Massachusetts) and the US National Museum's Smithsonian Institution (Washington, DC). This purchase included several very fine canfieldite and three spectacular stephanite specimens. Later in life, Ahlfeld sold many pieces to dealers including Jack

Arborescent group of parallel growth pseudomorphous crystals of acanthite after argentite from Colquechaca, Potosí department; 6.1 x 3.5 x 3 cm; Carnegie Museum of Natural History specimen (CM21532); Harold and Erica Van Pelt photo.

Young and Charles Key. His widow sold the rest, mainly to Terry Szenics, who sold the best pieces to a private US collector.

Frederick Canfield was an American mining engineer and passionate mineral collector. Although his mining practice was largely confined to New Jersey, Canfield visited Colquechaca twice in the 1880s and received literally hundreds of fine silver minerals. In his own words,

> *The writer* [Canfield] *saw collections of ores, from the various mining districts, in every large town that he visited. He got his best specimen of pyrargyrite in La Paz, but it came from Colquechaca. The type crystals of argyrodite were secured in Potosí, while the largest mass of this mineral was obtained in Sucre from the owner of a mine in Porco. Many similar cases could be given, but this is a sufficient explanation how any particular specimen could be found in La Paz. This is negative evidence. When in Colquechaca in 1887, the writer was presented with a specimen of canfieldite by the manager and part owner of a mine, in his office, at the entrance of the main adit of the mine, so there can be no question as to its locality. The principal crystal is much smaller than the one in the Brush Collection, but otherwise its appearance is identical with the type crystal, even to the line of twinning that is seen on the faces of the rhombic dodecahedron. As there is no direct evidence that this mineral has been found at any place besides Colquechaca, it is but reasonable to conclude that the type crystal came from the mines in that place. The specimen obtained by the writer at the mine in Colquechaca is a group of crystals of pyrargyrite with wire silver. Many crystals of*

Upper left: *Argyrodite; 6 cm wide; Bill and Carol Smith collection; Jeff Scovil photo*

Upper right: *A sign over the road into Colquechaca reads "Welcome to the Ruby Silver Capital of the World." Jaroslav Hyršl photo, 1997*

Lower: *Silver from Colquechaca; 6.8 cm tall; Frank and Kathryn Dzubek collection; Jeff Scovil photo*

canfieldite are scattered over the pyrargyrite. Most of them show on the dodecahedral faces.

Acanthite

Acanthite is common from Colquechaca and is occasionally associated with wire silver. The mineral is typically blocky and dull, but the individual crystals occasionally reach dimensions of 5 centimeters. Spectacular examples are on display at the Carnegie Museum of Nature and the Harvard Mineral Museum.

Argyrodite

Although germanium was first characterized by work in 1886 on Freiberg argyrodite, Bolivia has supplied the majority of the quality specimens. Argyrodite is known from several localities in Bolivia, Porco in particular. At Colquechaca the vein system is rich in argyrodite, and Canfield brought back a large number of argyrodite samples that made it into both private and museum collections.

Colquechaca argyrodite specimens reportedly have crystals as large as 3 centimeters that typically take shape as distorted dodecahedra. Botryoidal crusts of argyrodite, however, are more common.

Canfieldite

In 1894, Mr. Jacobo Aillou presented the British Museum with two specimens as examples of red silver from the mines of the Colquechaca. Many of the pyrargyrite crystals were studded with what looked like cubo-octahedra of acanthite with a bronze patina. The "bronze acanthite" turned out to be canfieldite, a very rare silver-tin sulfide later named for the famed American mineral collector.

Canfieldite is the "signature" mineral for Colquechaca, which is the only known locality for well-crystallized specimens. Crystals are generally pseudo-octahedra and reportedly can be as large as 1 centimeter. Most crystals are smaller and are usually tarnished with a blue or purple tint, giving specimens a characteristic bronze color. Colquechaca canfieldite commonly forms spinel-law twins.

Canfieldite was probably known for several decades from Colquechaca (and Porco) before it was actually described. A number of museums had canfieldite specimens erroneously labeled "brongniardite," often with the locality "Potosí," but the mines at Cerro Rico did not produce canfieldite, so these specimens are probably from Colquechaca.

Proustite

Proustite is much rarer than pyrargyrite at Colquechaca, but dozens of quality specimens grace museum and private collections. The crystals are

Upper left: *Canfieldite from the Veta Gallofa vein in Colquechaca; 2.2 cm wide; Terry Wallace collection; Jesse La Plante photo*

Upper right: *Stephanite; 3 cm tall; Terry Wallace collection; Jesse La Plante photo*

Lower: *Pyrargyrite; 4.5 cm tall; Bill and Carol Smith collection; Jeff Scovil photo*

rarely larger than 1.5 centimeter and are typically clusters on vug-filling white quartz.

Pyrargyrite

Pyrargyrite was the principle ore mineral at Colquechaca and occurred in such abundance that its specimen value was probably not fully appreciated by mine management during the heyday of production. Canfield's visits to the district netted him large numbers of specimens that were dispersed to American and European collections. Nearly every major mineral museum catalogue lists significant acquisitions from Colquechaca at the turn of the twentieth century. Some of the best specimens contain groups of pyrargyrite crystals in parallel growth with individuals reaching 5 centimeters in diameter and 12 centimeters in length.

In an 1891 memo entitled, "Fred Canfield's private collection at Dover—minerals that impressed me," Washington Roebling listed some 40 specimens, making notes about many. After pyrargyrite, he wrote, "Bolivia—hundreds of them."

Silver

Silver wires on acanthite were common at Colquechaca. Typically, the wires are spongy and most commonly have cross sections no greater than a few millimeters; however, thick striated crystals to 1 centimeter in diameter and 6 centimeters long growing from large acanthite crystals were once on display in the Banco Minero collection in La Paz.

Stephanite

Stephanite is common at Colquechaca and is typically found as pseudo-hexagonal roses. The largest crystals exceed 4 centimeters across and are commonly associated with pyrargyrite. Single elongated crystals to 2 centimeters are also known to have occurred.

A number of other silver minerals have been reported from Colquechaca, but in general, they are not of the quality of those listed previously. These include argentopyrite, dyscrasite, miargyrite, and polybasite.

Conclusions

Although Colquechaca is largely deserted today, a small cadre of miners scratch out a living cobbing ore from ancient pillars and stopes. The likelihood of a collectible silver mineral specimen being found is quite small, but that does not diminish the great history of Colquechaca and its extraordinary presence in fine mineral collections. The mine's pyrargyrite and canfieldite are among the world's finest and are coveted by collectors worldwide.

Thin Air and Coca Leaves

*by **Alfredo Petrov***

Wide-brimmed hats are the order of the day in the Bolivian highlands, where there is less atmosphere to protect human skin from the damaging exposure to sunlight.

Hungarian architect and polyglot Bendegúz Nagy photographed this garlic merchant in the market in Potosí in January 2006. Camera in hand, Bendegúz enjoys traveling and has visited countries in Latin and South America, Africa, Europe, and the Middle East.

In casual conversation about Bolivia, two topics almost invariably come up: the cocaine trade and the high altitude. Both can negatively impact mineral dealers and field collectors, although their affects can be mitigated, and they are less problematic than the first-time visitor might suppose.

In Thin Air

Anyone from a sea level city who has spent time in Leadville, Colorado, America's highest city (at 3,100 meters elevation), may have experienced a mild case of altitude sickness. This disorder can be quite miserable; it is not simply a matter of being short of breath. The physiology of altitude sickness, known as *sorojchi* or *soroche* in Bolivia, is not well understood. It involves changes in blood chemistry, among other things. Some of the effects are caused by reduced air pressure at high altitudes, not just reduced amounts of oxygen. The sufferer feels exhausted and out of breath after even minor exertion, such as walking up a slight slope. Other symptoms include digestive problems, lack of appetite, insomnia, and often a sharp headache that can't be cured with aspirin. The headache is caused by cerebral edema. In extreme cases, pulmonary edema can occur as well. Not a few mountain climbers have died from pulmonary and/or cerebral edema, brought on by their high-altitude activities.

High altitude has other effects as well. In thinner air, the stalling speed of aircraft is higher; so planes land and take off in La Paz at considerably higher speeds than at sea level. Years ago, when there were still commercial flights into Potosí, passenger planes could arrive full but could only take off half full; so there were always some disappointed passengers who had to take a bus or train to Sucre, the nearest lower-elevation airport. Golf balls travel much further in Oruro and La Paz than at sea level! Thin air holds less heat; so there is more temperature variation between sun and shade—one can feel hot in the sun and shiver with cold in a shady spot just a few feet away. Sunny days can be warm even in midwinter (June-July), but temperatures quickly drop to freezing after sunset. The ultraviolet light from the sun is much stronger than at sea level; so sunburn can be severe after even relatively short exposure. Consequently, skin cancer is widespread, especially occurring on the lips. Protection from the fierce sunlight is the reason the indigenous women wear wide-brimmed hats. The stars are much brighter, too: you have not lived until you have seen the night sky

from the Altiplano. Finally, some people claim that sex is more intense at high altitudes.

Leadville, Colorado, is more than 500 meters lower than the lowest place on the Altiplano, and some of the interesting Bolivian collecting localities are at over 5,000 meters elevation; so the high altitudes are understandably a cause for concern among potential travelers to Bolivia.

Altitude sickness is ideally prevented by staying a week at a moderate 2,500 to 3,000 meters' altitude, for example, in New Mexico, Colorado, or in the Bolivian cities of Cochabamba or Sucre, before heading up to the Altiplano. Such a stay allows the body to acclimatize and produce more red blood cells. A faster option is to spend the first days in Bolivia sleeping at a lower-elevation city and taking day trips to the highlands, such as from Cochabamba to Oruro or from Sucre to Potosi. Unfortunately, most of the other mining communities in the Altiplano are too far away for day trips.

Dehydration exacerbates altitude sickness, which is therefore mitigated by drinking plenty of fluids. Some doctors also recommend eating bananas to compensate for potassium loss. Medicines that can ameliorate some of the symptoms; Diamox is the best known of these. In cases of life-threatening edema, the only cure is to move immediately to a lower altitude. Native Bolivians have been familiar with altitude sickness for millennia; they swear by coca leaves for both prevention and cure.

Coca

Coca, *Erythroxylum coca*, grows 2 to 3 meters high, on the lower eastern slopes and foothills of the Andes, mainly in Bolivia, Peru and Colombia. Indigenous farmers have been cultivating the plant for at least 3,000 years, as demonstrated by coca found in mummies. Use of the plant in the Andes enjoys a tradition far longer than that of tea and coffee in our culture. Although it is grown in the more tropical lower elevations of the Andes, the principal consumers are the highland natives, who chew cheekfuls of dry coca leaves every day. The powerful drug. It was thought to sharpen the mind, and consumption of cocaine became popular among the intelligentsia. Cocaine even became an ingredient in the quintessential American drink, Coca Cola. In the twentieth century, as society became

Left: *Along with coca leaves, locals chew an alkaline paste, known as "llijta," many forms of which are found in Bolivian markets.*

Right: *Coca leaves are readily available in the markets of Bolivia.*

Both photos are by Frank de Wit

miners almost universally chew coca leaves, claiming that they would be incapable of hard physical labor at high altitudes without it.

The leaf contains tiny amounts of at least 10 or 11 different alkaloids, including cocaine and nicotine, and functions as both a stimulant and an anesthetic. Among other effects, it numbs the mouth and stomach, thereby reducing the sensation of hunger, and enabling miners to work longer hours underground without food. It provides significant nutrients to highland people, who consume few other green vegetables. To facilitate the dissolution of the alkaloids in saliva, coca leaves are usually chewed together with an alkaline paste called *llijta*, made from the ashes of cacti or quinoa, which contributes a few more minerals to the diet. White folks in Bolivia commonly consider llijta dirty and prefer to chew coca with a pinch of alkaline but nonnutritious bicarbonate of soda. Work is interrupted for coca breaks; the sharing of leaves and llijta is ceremonial, not unlike teatime in Great Britain.

Foreigners in Bolivia generally prefer to consume their coca as tea, and hotels and restaurants invariably recommend coca tea to lowland guests suffering from altitude sickness. Most of the time the tea at these establishments is so weak that any perceived improvement in one's mood can hardly be more than the placebo effect. Nevertheless, Bolivian government officials ceremoniously offer coca tea to visiting foreign dignitaries, including the Pope and the Queen of Spain.

The "War" on Drugs

In time, the coca plant has become a target of the War on Drugs. In the nineteenth century, Europeans and Americans discovered that cocaine, 0.8 weight percent of which is present in dry coca leaves, could be concentrated and made into a very much more more conscious of the dangers of addiction, cocaine use was restricted. It was dropped from the formula of Coca Cola about 20 years after the drink first hit the market, although cocaine-free leaves are still used in it today as a flavoring. In the mid-twentieth century, cocaine was classified as a narcotic, and its use was severely prohibited. Bolivia and the other coca-producing Andean countries began to suffer political pressure from abroad to reduce the cultivation of coca bushes.

Economic dependence on the United States has forced the Bolivian government to at least pay lip service to the idea of coca eradication, but the Bolivian populace has long resented intrusion by foreign powers into their domestic affairs. Farmers fear the loss of income from growing coca. Miners fear the loss of their traditional stimulant. Everyone fears the violent drug syndicates that control the international cocaine traffic.

Lest anyone get the erroneous impression that cocaine is legal in Bolivia, it is most definitely not! The leaves are legal, for chewing or tea, but possession of any paste or powder concentrated from those leaves will earn you a long vacation in a Bolivian prison.

The drug trade impacts mineral collectors as well: Bolivian mineral dealers cannot clean their minerals with simple chemicals such as hydrochloric acid, the possession of which is prohibited because it is one of the chemicals used to purify cocaine. And be forewarned that boxes of rocks carried or shipped out of Bolivia will be checked for drugs by police and customs officers, upon both leaving Bolivia and entering the USA. Packages are searched more often and more thoroughly than those from non-drug producing parts of the world; so there is a higher risk of damage from handling by inspectors. Visitors are well advised to use common sense when visiting this fascinating country.

Vivianite: Fleeting Beauty ...

by ***Alfredo Petrov****, who is variously living in Cochabamba, Peekskill (N.Y.), Tokyo, and southern California*

Most mineral collectors have, at one time or another, drooled over a gemmy green vivianite from Bolivia or elsewhere in the world but have hesitated to buy it, having heard that vivianite crystals are unstable and will eventually turn black and fall apart, especially when exposed to bright light. Perhaps, on the other hand, a collector went ahead and purchased a fine crystal in spite of any misgivings because some dealer made assurances that *this* fine and important specimen came from a locality that produces stable vivianite.

I have heard of people being sold vivianite that was alleged to be stable for a variety of reasons such as something present in the matrix of the specimen; or the absence of something in the matrix; or lack of oxygen at the high elevations at which the crystal grew; or lack of bacteria at that high elevation. Dealers also try to convince collectors that they, personally, will not have a problem with vivianite because they live in a sufficiently dry region or in a sufficiently humid one, or at a high altitude with low atmospheric oxygen.

A 1.8 cm tall Bolivian vivianite crystal; Michelle Paradise collection; Jeff Scovil photo

The bad news is that all vivianite from all localities suffers from the same type of instability; the good news is that its decomposition can be easily prevented. Before we discuss how to preserve vivianite specimens, however, we should understand the mechanism behind their self-destruction.

My favorite explanation for this phenomenon came from a couple of young counterculture types whom I met at the Tucson gem and mineral show. They had bought several Bolivian vivianite crystals from me at the prior year's show and had come back for more. "The last ones we got turned black already," they gleefully told me. "Did you leave them in a sunny window?", I asked, "You mustn't expose them to bright light." As it turned out, they *wanted* their vivianites to turn black and were delighted when the hoped-for phenomenon happened. From their point of view, vivianite gradually gets darker by absorbing the evil energy from its surroundings. After it turns black, one throws it (and all of the absorbed evil) away and buys a fresh one.

I can, however, offer an alternative explanation that involves a little more chemistry. Vivianite is a phosphate containing divalent iron and a lot of water: $Fe^{2+}_3(PO_4)_2 \cdot 8(H_2O)$. It crystallizes in several types of environments, which ought to be described before getting into the crystal chemistry.

One of these environments is mud and another is low-oxygen sedimentary rocks, usually in the reducing environment associated with fossil organic matter. The world's largest vivianite crystals (more than a meter long) occur, in muds in Cameroon. Aesthetic crystal sprays in fossil shells are found on the Kerch Peninsula in the Black Sea. And vivianite occurs in concretions found in lake bed and seafloor muds at several localities in Japan and in various US localities including Florida, New Jersey, and Contra Costa County, California. Buried bones and teeth are sometimes replaced by vivianite.

Vivianite also occurs in hydrothermal sulfide ore veins. The famous Bolivian vivianite forms in this environment, as does vivianite from Trepca in Kosovo, the Blackbird Mine in Idaho, the Ashio Mine in Japan, and other mines in Peru, the United Kingdom (Cornwall), the United States (Colorado), and elsewhere. Hydrothermal veins are the environment for splendid crystal specimens exhibiting classic gemmy green color.

Phosphate-rich pegmatites such as those in the Black Hills of South Dakota, and Brazil are a relatively minor source of vivianite. Other vivianite environments, of no importance to collectors, include rocks altered by coal mine fires and smelter slags.

The late Fred Pough observed that sedimentary

... And An Absorber of Evil?

vivianite changed color and flaked apart faster than hydrothermal vivianite. I have never timed the decomposition but tend to agree with him. Trace elements substituting for iron or phosphorous in the crystal lattice may be responsible for the more rapid decay of sedimentary vivianite.

Now, let's take a somewhat simplified look at the self-destruction mechanism. When photons (particles of light energy) enter a transparent vivianite crystal, they can knock a proton (a hydrogen nucleus) out of one of the eight water molecules, converting it into a hydroxyl ion (OH), which has a negative charge. This extra negative charge is balanced through the oxidation of one iron atom, whose valence state changes from 2^+ (ferrous) to 3^+ (ferric). Notice that no interaction with the atmosphere is involved: no oxygen entered the structure from outside, and no water molecule escaped to the outside.

Vivianite from Huanuni, Dalence province, Oruro department; the crystal is 4.8 cm tall; Mike Bergmann specimen; Jeff Scovil photo

The liberated hydrogen (the proton) migrates easily out of the crystal, leaving behind, not vivianite, but metavivianite, $Fe_2^{2+}Fe^{3+}(PO_4)_2(OH) \cdot 7(H_2O)$. The alteration is accompanied by a progressive color change; in fact, pure, fresh vivianite is colorless! A minor amount of light-induced oxidation rapidly (in a few minutes) changes the color to a brilliant transparent green. I was once offered a colorless vivianite crystal, which was wrapped in the world's filthiest greasy rag, while I was underground at the Morococala Mine (page 32). The miner who showed it to me set it out on his patio table under the burning tropical sun, and in fifteen minutes the crystals had turned bright green!

Further oxidation changes the color to deep emerald green before a strong pleochroism sets in and the color in transmitted light becomes cobalt blue when the crystal is rotated to a certain angle. Eventually the entire crystal turns deep blue and finally bluish black. An historical aside: powdered vivianite was, at one time, used in blue oil paint, with the color becoming more intense as time went on. In contrast, powered azurite in paint tends to alter slowly into malachite; hence, the green-tinged sky in some old paintings.

Theoretically, if another proton gets knocked out of the molecule, metavivianite alters to ferrostrunzite, $Fe^{2+}Fe_2^{3+}(PO_4)_2(OH)_2 \cdot 6(H_2O)$. Knocking out yet one more proton would leave $Fe^{3+}\ Fe_2^{3+}(PO_4)_2(OH)_3 \cdot 5(H_2O)$ or $Fe_3^{3+}(PO_4)_2(OH)_3 \cdot 5(H_2O)$, which correspond to the species ferristrunzite and santabarbaraite. No further reaction can take place because at this point there is no divalent iron left to oxidize. Since santabarbaraite forms pseudomorphs after vivianite, it would seem that these reactions can occur in Nature, but light may or may not play a role in the formation of these crystals.

We don't know which wavelengths of light are responsible for the reaction in vivianite. Many people assume that blue and ultraviolet light, the more energetic wavelengths, must be responsible, but this is not necessarily true. In the case of realgar, the green wavelengths of light have been shown to be the destructive agent, not blue or ultraviolet.

Since light causes vivianite to alter, the only thing a collector can do to preserve the diaphaneity and color of pristine vivianite crystals is keep them in the dark. That doesn't mean shade; it means in a closed cotton-lined box in a drawer, not a glass case. Let your casual visitors drool over your green tourmaline, emerald, and forsterite, and reserve the vivianite for short-term viewing by your connoisseur friends—it will take a long time for a few minutes of occasional indoor exposure to noticeably change color. Or take a cue from my counterculture friends and replace the specimen once it has absorbed its fill of evil (light) energy.

The Viloco Tin District

by
Jaroslav Hyršl
and
Robert B. Cook

Quartz and cassiterite from the Viloco district, near Araca in Loayza province, La Paz (repaired); 5 cm tall; Dave Bunk Minerals specimen (ex Jim and Dawn Minette); Jesse La Plante photo

World-Class Cassiterite

World traveller and mineral dealer Rock Currier (yes, his real name) once said that the road to the Viloco was the most scenic he had ever driven. It would be difficult to argue with him. Winding through staggering peaks flanked with glaciers, lakes, moraines, grazing llamas, and colorfully dressed people, in the fierce sunlight that is typical of tropical latitudes, is indeed a memorable driving experience.

Viloco is a tin mine in a district of the same name located near the old mining community of Araca in Loayza Province, La Paz department; the mine is one of several sometimes referred to as "Araca" or the "Araca Mine." The district occupies an area of approximately 5 square kilometers along the western margin of the Quimsa Cruz Batholith, approximately 7 hours by road south of La Paz. Turneaure and Welter (1947) described the geology of the veins.

For the past 80 or more years, these mines, which were developed on a series of veins formerly exploited by La Empresa Minera de Araca, have been producing what are undoubtedly among the world's best cassiterite specimens. The major mine is currently in operation, but top finds are limited; still, over its productive life, this prolific group of mines has provided most of the world's major museums and private mineral collectors with exceptional specimens, some distinguished by their large size.

Cassiterite with quartz from the Viloco district, near Araca in Loayza province, La Paz department; 3.2 cm tall; Paul Stahl collection; Jeff Scovil photo

Cassiterite

Cassiterite is the principal ore mineral at Viloco, where it occurs as short prismatic crystals. Because of their beauty and perfection, Viloco or Araca cassiterite crystals have been described in detail by various authors including (Gordon, 1924; Heilmeir, 1930; Ahlfeld and Munos Reyes, 1943; and Cook, 1977). At least ten crystal forms have been described (Ahlfeld and Munoz Reyes, 1955), and it is typically distinguished by twinning on {011}. Crystals generally vary in size from 1 to 3 centimeters, although some as large as 8 centimeters on edge have been found. Groups of 2- to 3- centimeter, lustrous cassiterite crystals aggregated on matrix in specimens weighing up to about 100 kilograms are known from this locality. Crystals are commonly sharp but sometimes exhibit rounded edges, possibly a result of natural etching. Viloco cassiterite tends to be dark coffee brown to almost black and highly lustrous. Some veins have produced lighter brown, yellow, and even colorless crystals, the latter containing very little iron. Color-zoned crystals with black cores and pale surfaces occur, as do specimens with crystallographically oriented dark and pale zones. Crystals coated with siderite are often anomalously light colored beneath. Colorless crystal sections can be faceted; in fact, Viloco rough has netted many fine cut stones, which at times reach 5 carats and in rare cases can weight as much as 15 carats (Hyršl, 2002). Most crystals, however, are included with micro acicular schorl crystals; so large, clean-cut specimens are relatively rare.

Other Minerals

Arsenopyrite is common and forms radial groups of flat prismatic crystals. Beautiful specimens of cassiterite crystals to about 2 centimeters, in association with arsenopyrite crystals to about 5 centimeters, were found in the 1970s and again in 2000. Arsenopyrite was deposited before the cassiterite, and occasionally, transparent cassiterite is found overgrowing arsenopyrite. Small **sphalerite** crystals are also common, and in rare instances black crystals exceeding 1 centimeter long have been found; some of these are twinned. Viloco sphalerite and cassiterite are similar in appearance. Small **pyrrhotite** crystals on quartz were found in close association with vuggy cassiterite-rich veins in the San Antonio Mine in the 1970s.

Molybdenite is common in some veins, but it only forms flakes to about 2 centimeters—no crystals. **Ferberite** is rare at Viloco, occurring as

***Upper**: Quartz from the Mocoya Mine near Araca in Loayza Province, La Paz department; 11.6 cm wide; George Witters collection; Jeff Scovil photo*

***Lower**: Cassiterite and arsenopyrite; 10 cm wide; Jaroslav Hyršl collection and photo*

Upper left: *Tourmaline and cassiterite; the field of view is 6.5 cm tall; Jaroslav Hyršl collection and photo*

Upper right: *Cassiterite and quartz from the Viloco district; 8.8 cm tall; Paul Stahl collection; Jeff Scovil photo*

Lower: *Apatite on siderite with cassiterite; 5.5 cm wide; Jaroslav Hyršl collection and photo*

druses of lustrous crystals reaching 3 centimeters on quartz. Native **tin** occurs at Viloco, although very rarely. It has been identified once by Alfredo Petrov, who noted it on an old specimen. That specimen consists of numerous cassiterite crystals, one of which is partly covered with silvery native tin that looks like a dendritic manganese oxide.

Quartz is commonly associated with cassiterite from Viloco. It forms simple prismatic crystals, which can exceed 3 centimeters in length. On rare occasions, sceptered quartz has been found. For nearly a decade, the nearby Rosario de Asiento gold-tin mine has been producing smoky quartz in druses and in crystals up to about 10 centimeters long. Some exhibit dark phantoms.

Siderite occurs as small yellow-brown rhombohedra or, more rarely, lenticular crystals. Well-formed simple prismatic **apatite** (probably fluorapatite) crystals are found, but are very rare. They are associated with siderite and can reach lengths of 1 centimeter.

Schorl is another common mineral at Viloco. It was one of the earliest minerals deposited and forms fine gray to black needles, which can be more than 1 centimeter long. As mentioned above, schorl is commonly found included in cassiterite. Ahlfeld and Munoz Reyes (1955) describe a vein in Viloco that consisted of some 80 percent schorl in association with wolframite group minerals and arsenopyrite. In late 2001, cassiterite druses in association with abundant gray tourmaline needles up to 2 centimeters long were found. Very rarely, small rosettes of green **clinochlore** occur on black cassiterite crystals.

Visiting

In recent years, the Viloco area has become a staging point for trekkers and mountain climbers, an unusual example of a remote mining camp accustomed to seeing tourists other than mineral collectors.

The Viloco mine dumps are there for the collecting, but the work is difficult, primarily because of the high altitude. Inactive parts of some of the older, higher mines have been scavenged for small zones of high-grade ore. The miners who worked those zones collected small single cassiterite crystals along with the ore. Although reluctant to show their accumulations of cassiterite, miners will sell these crystals on occasion.

Native Copper and Unusual Pseudomorphs from Corocoro

by ***Jaroslav Hyršl*** *and* ***Alfredo Petrov***

Corocoro, a very unusual copper deposit located on the Altiplano about 130 kilometers southwest of La Paz, is famous among collectors for native copper pseudomorphs after aragonite, which are unique.

History

Michigan's Keeweenaw Peninsula and Corocoro are the only significant copper deposits in the world in which the native metal occurs abundantly as the dominant ore mineral. Corocoro was well-known to ancient Native American civilizations, a fact born out by the pre-Columbian copper tools and other objects housed in many archeological museums. Of course, there are no written records from this period, and subsequent activity has damaged or destroyed most of the remains of ancient mining efforts.

The first written report about Corocoro dates from 1628 and mentions Indians panning copper in the Guallatiri River. Deep mining started at the beginning of the eighteenth century, and the deposit was divided into numerous small mines. Rumors of an ancient cuprified miner may date from this period. The Chilean company Compania Minera Boliviana brought together the more central properties of the district in 1873; Corocoro United Copper Mines, a French company, purchased the remainder in 1909. Compania Americana Smelting Boliviana took over the deposit in 1937, and with the revolution of 1952 the mines were nationalized and became part of COMIBOL. The organization could not make a profit from its Corocoro branch and abandoned it, leaving behind great heaps of rusting machinery and ore cars. A few miners' cooperatives struggle on, digging with hand tools in small near-surface adits and "rat holes."

At some sites, thousands upon thousands of the famous pseudomorphs (albeit mostly malformed or incomplete) are the actual ore, concentrated merely by pushing a wheelbarrow load one kilometer down to a creek to wash the mud off! Those nice thumbnails one sees for sale at mineral shows are not nearly as rare as one might like to believe.

Throughout its productive history, Corocoro was by far the most significant copper district in Bolivia, with total production estimated to be 350,000 tonnes of the metal (Entwistle and Gouin, 1955) or perhaps even 450,000 tonnes (Ahlfeld and Schneider-Scherbina, 1964). The deepest mines of the district reached a depth of about 450 meters (1,475 feet).

Geology

Two main types of clastic sedimentary rocks occur in the Corocoro area. The two rock types are

Copper pseudomorphs after aragonite from Corocoro in Pacajes province, La Paz department; 1.3 and 1.5 cm tall; Jim McEwen collection; Jeff Scovil photo

separated by the 70-kilometer-long, very distinct, north-south trending Corocoro fault. The eastern beds, known as the *Ramos*, are composed of red shales, sandstones, and a few conglomerate beds. The younger, western unit, called the *Vetas*, is made up of sandstones with subsidiary arkosic rocks and minor conglomerates. (*Vetas* means "veins" in Spanish, but these are not hydrothermal veins in the usual sense.) Both the Ramos and Vetas probably date to the Pliocene and represent typical continental sediments from a semiarid environment, as evidenced by raindrop prints, mud cracks, worm tracks, and evaporite rocks. Fossil wood, twigs, coprolites, and even worm burrows replaced by chalcocite, native copper, or cuprite are not particularly rare.

Mineralization occurs in both units, but only in certain beds along the Corocoro fault. Native copper is characteristic of the Ramos unit, which has an oxidation zone at the near surface part of the deposit. Chalcocite occurs only in the deeper part of the Vetas in arkose and some sandstones, while the upper part contains native copper and oxidation products. Bonanzas of metallic copper were near the fault.

The generally accepted theory of mineralization was proposed by Adrien Berton (1937), who spent 21 years in Corocoro as a geologist. Berton supposed an epigenetic origin with a deep magmatic source. Precipitation of native copper or chalcocite was controlled by the presence of Fe_2O_3 in the sediments. Where Fe^{3+} was present, native copper originated. In the absence of Fe^{3+}, chalcocite formed. This theoretical epigenetic origin was accepted by Ahlfeld (1953), who studied not only Corocoro but also several smaller similar copper deposits on the Altiplano.

***Right**: Native copper from Corocoro; 7.6 cm tall; Tom Hughes collection; Jeff Scovil photo*

***Below:** Miners work some of the numerous "rat holes" that access the Corocoro copper deposit. Frank de Wit photo (2009)*

Minerals

Native **copper** is the most common mineral in Corocoro. It generally forms as the cementing matrix of sandstones or sheets or plates filling fractures in the sediments. Large sheets of copper have been found that are up to 20 centimeters thick with surface areas as great as 12 square meters! Crystallized specimens are relatively rare but usually occur as arborescent aggregates of deformed dodecahedra and octahedra. Copper, containing 10 to 12 weight percent arsenic (the variety whitneyite), is in rare cases found in association with **domeykite** (Cu_3As).

The pseudomorphs of **copper after aragonite** were first described by Chilean mineralogist Ignacio Domeyko (1880). They come in two types: 1) relatively simple pseudohexagonal crystals, and 2) "balls" composed of equant interpenetrating crystals. These pseudomorphs can reach 6 centimeters in diameter, but generally measure only 1 to 2 centimeters. Copper-bearing solutions replaced the

***Upper left**: Corocoro pseudomorphs; Frank de Wit photo (2009)*

***Upper right**: Cuprite from Corocoro; 3.5 cm wide; Jaroslav Hyršl specimen and photo*

***Middle**: Piles of oxidized copper wait to be hauled away. Jaroslav Hyršl photo*

***Lower**: Calcite and limonite after halite from a locality about 12 kilometers outside of Corocoro; 3 cm on edge; Anne Cook collection; Jesse La Plante photo*

aragonite from the outside in, and many specimens are only partial replacements. The degree of replacement can be estimated by measuring a specimen's density. Most of the pseudomorphs are coated with cuprite and green malachite-stained clay when they are first recovered. The bright copper-red metallic specimens on the market have generally been cleaned with acid.

Striking pseudomorphs of calcite and iron oxides replacing halite occur in the vicinity of Corocoro. The rusty color of these specimens can make them appear, at first glance, as copper replacements, but their weight (too light to be copper) and a look with a loop gives them away.

Native **silver** is, like copper, found as a cementation product, and rich occurrences of silver are mentioned in old reports. The authors have seen a few old specimens of silver as the cementing agent in sandstone, but such specimens are rare today.

Rich **domeykite** mineralization with minor **algodonite** is known from the deep parts of the Vetas, often occurring with celestine. Interestingly, the chalcocite and domeykite mineralizations are sharply separated from each other. Rare occurrences of domeykite in the oxidation zone led to the formation of small crystals of secondary **olivenite** and chalcophyllite—rare examples of interesting Bolivian secondary minerals.

Chalcocite is very abundant in Corocoro, but no crystals have been found; instead, it is a component of the sandstones and arkoses of the Vetas. At Corocoro, chalcocite has a low silver content (0.006 weight percent Ag). Rare samples of chalcocite replacing wood, however, are interesting from the mineralogical point of view. Amstutz (1981) described a copper pseudomorph after aragonite in which chalcocite partly replaced copper from the surface inwards.

Celestine and its barium-bearing variety barytocelestite were at one time common in

Corocoro. Transparent crystals can reach lengths of 2 centimeters and are white, blue, or pink. Another celestine variety exhibits a fibrous habit like that of alabaster and is found in association with native copper and domeykite in the Vetas Mine.

Gypsum is common in many sediments of the Corocoro area, imparting the white color to the Cerros Blancos (White Hills) near the town. Gypsum crystals are common on fractures and are sometimes found intergrown or included with native copper.

Cuprite is the most common oxide mineral in the oxidation zone, although some would argue that native copper is in the oxidation zone as well. On rare occasions, druses of translucent crystals up to 6 millimeters were found in fractures, and druses of tiny cuprite crystals commonly coat specimens of crystallized copper and the pseudomorphs. An unusual type of pseudomorph—cuprite after aragonite—up to 5 centimeters long, was found by the authors. These two-stage pseudomorphs have rounded edges and are fractured, having expanded in volume as the copper, which had initially replaced aragonite (replacement pseudomorph), altered to cuprite (alteration pseudomorph). The alteration probably took place as a result of the copper's having oxidized after the water table sank.

The area's best **malachite** specimens, thick crusts or stalactites, are from the Marcocoya Mine. Malachite was abundant in pre-Colonial times and was very popular in ancient Indian jewelry. Corocoro malachite has been found in many archaeological sites, including Tiahuanaco.

***Upper left**: Celestine from Corocoro; the field of view is 5 cm wide; Jaroslav Hyršl collection and photo*

***Upper right**: Copper and cuprite from Corocoro; 8.2 cm tall; Al and Sue Liebetrau collection (ex Marty Zinn); Jeff Scovil photo*

***Left**: Aragonite from Corocoro; 3 cm tall; Jaroslav Hyršl collection and photo*

***Inset**: Malachite on azurite; the left spray is 1.6 mm long (6.7x); sDan Behnke collection and photo*

The Cerro Sapo Sodalite Deposit: A Pre-Columbian Gem Mine

by ***Alfredo Petrov***

Polished sodalite set in sterling silver with massive pieces of sodalite; the sodalite originated on Cerro Sapo, in Cochabamba department, Bolivia. Alfredo Petrov collection; Jeff Scovil photo

Discovery and Rediscovery

In the late nineteenth century when legitimate archaeologists (as opposed to grave robbers and other unscrupulous treasure-hunters), started studying the numerous ruins left by pre-Columbian cultures in the Andes, they discovered abundant cylindrical blue sodalite beads and (more rarely) small sodalite figurines of gods and llamas. Pre-Columbian sodalite has been found as far north as Quito in Ecuador; at the ruins of Ica, Ancón, and others in Peru; at Tiahuanaco and many other sites in Bolivia; in northern Chile; and as far south as Tucuman in Argentina. The Tiahuanaco (approximately 650 to 1050 CE) and later the Incas were the most prolific producers of these artifacts. However, sodalite bead use came to a sudden halt when the Spanish *conquistadors* destroyed the Inca Empire in the early sixteenth century and turned the native populations into forced laborers for the extraction of precious metals. By 1572, the source deposit for the sodalite was forgotten.

Some sixty years later in 1635, the Jesuits showed up in Bolivia and built the little-known Plazuela mission at Sacambaya in the deep Ayopaya river valley to serve the spiritual needs of the far-flung and not very productive Cocapata placer gold district. They rediscovered the sodalite deposit and for more than a hundred years, shipped blocks of wonderful sodalite, commonly misidentified as "lapis lazuli," back to Europe for use in altars and columns. Some of these creations still adorn European buildings. The Jesuits were secretive about the provenance of their "lapis," and the old-world customers had no idea where to get more after the Jesuits were expelled from South America in 1789.

The fact that no massive Andean sodalite deposits were known to late nineteenth and early twentieth century archaeologists was commonly offered as "proof" that ancient South American Indians had traded with cultures on other continents! Andean sodalite, however, is mineralogically unique in that it occurs with ankerite and barite rather than the usual nepheline and feldspar. These associations strongly indicate that the material came from a single source.

The colorful story of the modern rediscovery of Bolivian sodalite begins with Edgar Sanders, a Russian-born British adventurer and scam artist and with German geologist Friedrich Ahlfeld, for whom the mineral ahlfeldite is named. Sanders came to Bolivia with two European partners in 1925 to hunt for gold treasures allegedly left behind by the Jesuits. Somehow Sanders got hold of a document indicating that a treasure was buried in an old mine tunnel near the ruins of the Plazuela mission. (There are still occasional notices and classified ads in Cochabamba offering ancient treasure maps for sale, and one can only wonder why the sellers don't dig up the treasure themselves.) After some work, Sanders was unable to find the supposed treasure; so before returning to London, he buried his document in an opening that had been washed out of the banks of the Rio Kazu, a tributary of the Ayopaya.

In 1927, he raised £25,000 sterling by selling shares in his newly established treasure-hunting company. By 1928, he returned to the Plazuela ruins with several tonnes of mining equipment and 15 fellow adventurers. They set to work on the banks of the Rio Kazu and sank a square shaft. The mining crew's spirits soared when they "accidentally" hit Sander's old treasure map. By September, however, with the southern summer rainy season fast approaching, hope had faded.

Sanders tried to salvage his credibility by suggesting that they invite a scientist to join the expedition. After all, gold had been worked in nearby regions on a small scale during the Jesuit era, and a professional report stating that workable gold veins existed in the area would potentially start unlimited funds flowing from London. Austrian-born photographer Robert Gerstmann (1896 to about 1960) was along to photograph the expedition. Gerstmann recommended his friend and passionate fellow photographer, 35-year-old Friedrich Ahlfeld, who had not yet earned his Ph.D.

Ahlfeld smelled adventure and was happy to join the expedition, in part because he already suspected that a sodalite occurrence was somewhere nearby. He had previously seen a piece of sodalite in a Cochabamba museum labeled "Palca," the small town in this mostly roadless region of Bolivia that is nearest the deposit. (By the way, the name is not on modern maps because the town is known today as Independencia.) In addition, Ahlfeld's good friend Hans Block (another German geologist, better known as the owner of the Pacajake selenium mine) had found a sodalite-bearing outcrop in the area a few months earlier (1928). Although Block never published anything on the occurrence, it is quite

Lapis vs. Sodalite

"**Lapis lazuli**" is the trade name for a variety of the sodalite group mineral lazurite $(Na,Ca)_8Si_6Al_6O_{24}[(SO_4),S,Cl,(OH)]_2$. Lapidary-grade lapis is commonly a natural composite of lazurite and other minerals including pyrite (desirable), and calcite (not so desirable). **Sodalite** is a related to lazurite but is a distinct mineral species with the chemical formula $Na_8Al_6Si_6O_{24}Cl_2$.
A streak test easily distinguishes sodalite from lazurite/lapis lazuli: sodalite has a white streak/powder, lazurite a blue streak; so sodalite cannot be used as a pigment.

Alfredo Petrov is pictured standing in the "Eyes of the Incas," an early sodalite excavation on Cerro Sapo.
John Lucking photo

likely that he mentioned it to Ahlfeld.

An exploration team was formed consisting of Ahlfeld and three of the English treasure hunters. The team went east from the Jesuit ruins and on up the very steep slopes of Cerro Sapo (4,200 meters elevation). There they spent three weeks camping in an alpine meadow high above the cloud forest while Ahlfeld supposedly prospected for gold—his team members had not fully realized that he was far more interested in the exotic mineralogy of a sodalite deposit than in some meagerly gold-bearing, boring quartz veins.

Ahlfeld described his awe on discovering, in an almost impenetrable ravine on the northeastern slopes of Cerro Sapo, a "blue grotto," an ancient, hand-dug cavern in a great sodalite dike, "hundreds of meters long and branching inside the mountain, the wet walls shining a magnificent dark blue, an indescribable view, these shimmering halls in the damp cloud forest." (Neumann-Redlin, 2007 and personal communication with Fritz Berndt). "Hundreds of meters" was a bit of an exaggeration on Ahlfeld's part, since the longest tunnel actually measured 90 meters. He discovered nine ancient tunnels in all, one reaching a height of 5 meters. Three weeks of prospecting on Cerro Sapo revealed only nepheline syenite and a complex variety of other alkaline rocks (including the sodalite-bearing dike), but no traces of gold.

The bad news had reached the main camp on the Ayopaya River before Ahlfeld and his crew returned. The other treasure hunters felt they had been robbed of two dreams: the golden treasure and the prospect of additional minable gold veins. Perhaps some of them even suspected that Ahlfeld had devoted his time to the study of minerals other than gold. Consequently, a nasty mood hung in the air when Ahlfeld returned to camp, and only the intervention of his friend Gerstmann saved him from having his teeth knocked out.

Sanders remained convinced that there were rich gold deposits on Cerro Sapo. He and Ahlfeld had a heated argument, and Ahlfeld ended the row by angrily shouting, "I'll give £1,000 [sterling] to anyone who shows me a single workable gold vein on Cerro Sapo!" Ahlfeld then stealthily fled the camp and sought refuge at the head of the valley in the cozy Valenzuela Inn, where, he later admitted, he drowned his anger in two bottles of golden yellow Santa Rita wine.

Our debonair but scurrilous friend Edgar Sanders popped up again on the international scene more than twenty years later, during the Cold War, when he was arrested in Hungary. He was accused of being a western spy operating behind the iron curtain and was sentenced to 13 years in prison in 1950. It took an international effort, including the intervention of Winston Churchill, to gain him a pardon; he was out after serving only three years.

Mining Past and Present

When entering one of the ancient tunnels today, visitors can clearly see evidence that pre-Columbian miners worked the hard rock by fire setting: they lit fires against the rock face, then threw cold water on the hot rock to shatter it by thermal shock. Given its good cleavage in six (!) directions, the sodalite broke into mostly thumbnail-sized pieces; thus, pre-Columbian cultures only used sodalite to make small objects such as beads. These miners had no metal tools, only stones—rounded river cobbles of much harder rock types carried up from canyons far below the mine—for hammering and deer antlers (some friable remains are still present) for digging. Enrico Zappa found a now very friable ancient deer antler in the mine along with the skeleton of a teenage girl, the latter showing that the ancients probably did not share the current Bolivian miners' European-derived superstition forbidding women underground.

Several times after Ahlfeld's discovery, various small-scale Bolivian and European endeavors attempted to extract the sodalite using hand tools and

dynamite, carrying the blue product out in sacks loaded on the backs of mules. None of these ventures was economically successful until 1989, when the claims were purchased by the well-financed quarrying company Dimentional Sodalite, Ltd. Under the capable management of Italian-South African immigrant Enrico Zappa, Dimentional spent several years building the 30-kilometer-long road from Independencia to Cerro Sapo as well as a modern, well-equipped mining camp, several hundred meters downhill from the deposit.

Dimentional Sodalite brought in large diamond saws, which enabled them to make slabs and cubic-meter-sized blocks. Sodalite is now regularly exported to Italy, the United States, and the Middle East. The company also exports sodalite chips used in the manufacture of blue "marble" floors such as the one in the new terminal of the Düsseldorf airport in Germany.

Geology and Mineralogy

The sodalite occurrence on Cerro Sapo is not unique in the region, and several others, including a 40-centimeter-wide dike on Cerro San Cristobal, have been found since Ahlfeld's rediscovery. The authors have been studying a small galena-rich sodalite vein located on a steep slope close to Independencia, approximately 30 kilometers south of Cerro Sapo. The new find produces good-quality sodalite, but the vein is just 3 centimeters wide and is very difficult to access. The strike of the vein is the same as the one on Cerro Sapo, indicating that the appropriate geology extends over a wide region, so more veins and dikes probably exist.

The sodalite-bearing dike on Cerro Sapo is a unique igneous rock type: an ankerite-carbonatite, which is about 3 kilometers long, varies from 1 to 11 meters thick, and is generally less than 5 meters thick. It trends northwest to southeast and cuts through a fine-grained nepheline syenite intruding Paleozoic slates and through the large hornfels aureole surrounding the intrusion. The hornfels is extremely tough; hence the steep and rugged topography of Cerro Sapo.

The main constituents of the carbonatite dike are, in descending order, greenish white ankerite, deep blue sodalite, light grey to pink barite, white natrolite, dark red-brown hisingerite, pale brown siderite, silky pinkish dawsonite, reddish burbankite, minor accessory sulfides (sphalerite, galena, pyrite, and chalcopyrite), and numerous other species. The sodalite can exhibit large grain sizes, with individual cleavage faces up to 11 centimeters across. Pockets containing sodalite crystals are very rare, although crude euhedral crystals to a few centimeters are sometimes found embedded in solid ankerite, barite, or hisingerite. Other cavities do, however, yield

Upper: *Alfredo Petrov points out a single sodalite crystal in the quarry wall.*

Middle: *Quarry operator Alain uses his homemade wire saw to cut one-ton blocks of sodalite.*

Lower: *The sodalite mining camp (white building) is seen through clouds from the road to Cerro Sapo.*

All three photos were taken by John Lucking.

Pre-Columbian beads and rough all originating in Cerro Sapo; the largest bead is 2 cm tall; private collection; Jesse La Plante photo

some of the world's nicest ankerite crystal druses, which show an unusual pale green color and are associated with colorless analcime crystals and well-terminated pinkish white dawsonite. In addition, the occurrence of what are perhaps the world's most interesting **amesite** crystals, which occur as green, six-pointed stars(!) are a particular thrill for micromounters. Other rarities include light blue, radiating spheres of world-class **krasnovite**, along with well-crystallized **nordstrandite**, **daqingshanite-(Ce)**, and **goyazite**.

Rumors, Lies, and Misinformation

Rumors of sodalite occurrences in more distant parts of Bolivia are false. Reports of sodalite from near the small towns of Arani and Tarata in the Valle Alto de Cochabamba are based on finds of sodalite chips in the local soils that, in reality, were left by Inca-era lapidaries plying their craft in workshops that were located in the area. Stories of sodalite from Norte de Potosí and the Uyuni area were deliberately fabricated by highgraders selling material stolen from the Dimentional Sodalite claims.

Today, there are still rumors of gold occurring within the Cerro Sapo sodalite deposit, and some local natives are quite certain that the foreigners really want the blue rock for its gold content. These rumors may be left over from the old Jesuit treasure yarns and are reinforced by the glittery golden specks of pyrite and chalcopyrite sparsely disseminated in the carbonatite. The treasure we know we have for sure is the enduring blue color of the sodalite and the unexpected beauty of the tiny crystals sometimes found in a pocket.

WARNING for collectors and tourists: Travelers in Peru and Bolivia are sometimes offered "pre-Columbian" beads. These are frequently modern reproductions; however, when beads are genuinely pre-Columbian, removing them from their country of origin is illegal! The easiest way to distinguish ancient from modern beads is to look at the hole: Beads made with modern drills have cylindrical holes; ancient beads drilled with bows, sticks, and sand have conical holes. (Nobody has the patience to drill holes that way nowadays.)

Field of Stars
Six-fold sector twinning is responsible for the star-shaped terminations of these amesite crystals from Cerro Sapo. The terminations are 200 microns in diameter. The crystals are 500 microns long. Collection and photomicrograph Saul Krotki

The Anahí Mine

by ***Jaroslav Hyršl*** *and* ***Alfredo Petrov***

History

Bicolor purple-yellow quartz is by far the most popular Bolivian gemstone. Local jewelers refer to it as "bolivianita," after the country, although outside of Bolivia almost everyone calls it "ametrine," for the purple quartz variety "amethyst" and the yellow variety "citrine." Bolivian sellers allege that this gem was discovered in the seventeenth century, and they promote a romantic story involving a liaison between a Spanish conquistador and Anahí, a "Princess" of the Ayoreos, the primitive Native American tribe inhabiting the jungles of far eastern Santa Cruz department, where ametrine is found. There is no documentation for this legend, and dealers may have simply made it up for marketing purposes. In fact, there is no evidence that the Ayoreos had any interest in the quartz until they were offered cash and goods to work in the mines.

More credible sources claim that the enormous quartz dikes (or veins) that carry the ametrine were first noticed in the early twentieth century by British engineers scouting for a location to build a port on one of the lakes on the border between Bolivia and Brazil. The lake connects to water transportation on the river Paraguay.

The first recorded mining was carried out in the 1960s, mostly by Brazilians sneaking across the poorly controlled border into Bolivia. But the world was unaware of ametrine's unique color zones because the Brazilians sawed the stones and only used the amethyst sections for faceting! Bicolor gems began to appear sporadically on the market in the late 1970s, when the military dictators who then ruled Bolivia realized the deposit's potential and started mining it themselves. In the late 70s and early 80s, military officers sold the gem rough illegally, without paying taxes, to buyers in the Brazilian border town of Corumbá. These activities are on the long list of criminal charges for which ex-dictator General Garcia Meza is now sitting in prison. The Brazilians did not pass on the locality information; so for many years, ametrine was wrongly thought to be from Brazil or even to be synthetic.

Faceted quartz (variety ametrine) 82.35 carats The rough for this gem originated in the Anahí Mine, Puerto Suarez, Santa Cruz department. Harold Dibble collection Jeff Scovil photo

Thanks to democratization, the end of military rule, and the modernization of the mining law in Bolivia in 1989, the locality came to be exploited by a private Bolivian company, and the true location of the world's only commercial ametrine deposit

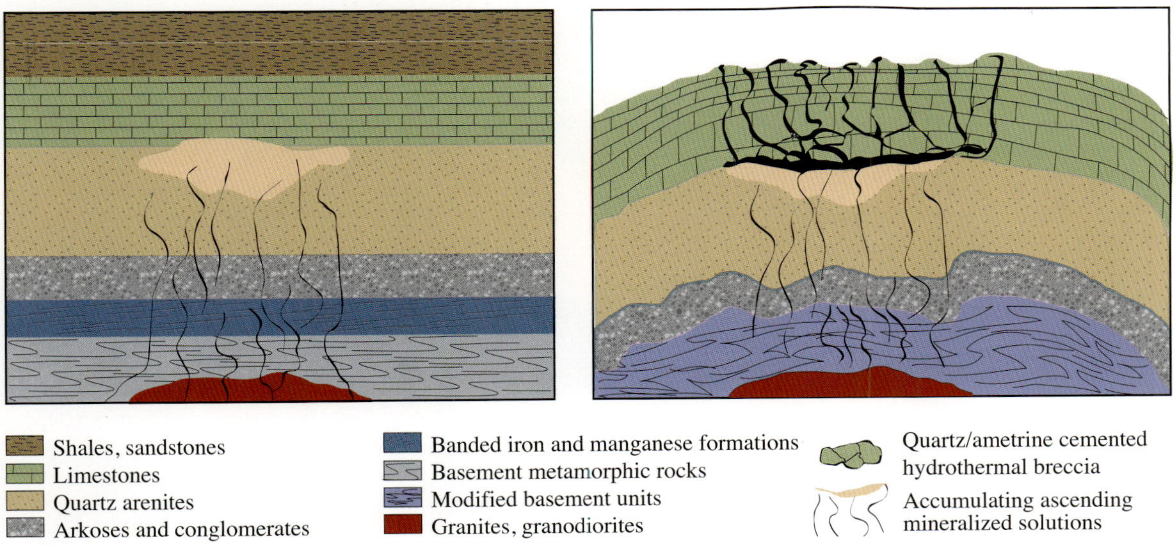

Schematic cross section of the formation of the Anahí ametrine mine after Vasconcelos et al. (1994)

became publicly known and was very thoroughly described by Vasconcelos et al. (1994). Vasconcelos' article convinced skeptics that ametrine was neither synthetic nor from Brazil, although crystals are still sometimes mistakenly labeled "Corumbá, Brazil." Since the details of the deposit have been brought to light, large quantities of both crystals and cut stones are easily available on the market.

Notice the rare {0001} pinacoid on this 6.5 cm wide quartz, variety ametrine, from the Anahí Mine. Kosnar collection; Brian Kosnar photo

The Locality

Ametrine comes mainly from the Anahí Mine near Lake La Gaiba on the Bolivian-Brazilian border in Santa Cruz department. Other mines a few kilometers away, but exploiting the same quartz dike, include Mina Pobre and Mina Ayoreíta, but the Anahí Mine (also spelled "Anai" and "Anay") is the only one with commercial quantities of good quality ametrine; the other mines yield only amethyst or amethyst-smoky combinations.

The range of hills comprising this mining district rise up out of the Pantanal, a large flat, often flooded jungle region in which travel by vehicle is possible only in the dry season. During the rainy season, travel to the mines is by canoe or air taxi. Incidentally, *Anay* is the Ayoreo word for "hand" and refers to the shape of a nearby lake that is the main source of water during the dry season. *Anai* and *Anahí* are derivations created by non-Ayoreo speakers.

The quartz veins are hosted in Cambrian dolomitic limestones, which are slightly silicified and form morphologically significant hills. The hydrothermal quartz veins are associated with long regional faults trending north-south. The limestones at the mine are strongly brecciated and the blocks are cemented by vein quartz. Large quartz vugs formed between the largest blocks. Some crystal-lined chambers can reach lengths of up to 5 meters, widths of 3 meters and heights of up to 5 meters and can be as voluminous as a Tokyo studio apartment.

The deposit originated as ascending silica-rich, moderately iron-enriched fluids, which created high pressure on the overlying sedimentary limestones. When the hydrostatic pressure became higher than the lithostatic pressure, the limestones broke abruptly and silica started to precipitate very rapidly between the blocks. The resulting ametrine crystals can be up to 40 centimeters long; the largest yield the best gem rough. Smaller crystals tend to be sharper, although they sometimes terminate with an uneven *c* face, a phenomenon which is extremely rare on quartz crystals from other localities. Some crystals also exhibit strong etching features on their surfaces.

The unique zoning in ametrine is very interesting, especially because the color boundaries are quite sharp. The major rhombohedral sectors are purple with strong Brazil-law twinning, while minor rhombohedral sections are yellow and not twinned. The best gem material is found at the connection of rhombohedral and prismatic faces. An electron microprobe study showed that the citrine zones have a much higher iron content (68–125 ppm) than the amethyst (19–40 ppm) and colorless (~17 ppm) zones. Ametrine has been synthesized by Russian scientists from iron-, potassium-, and carbonate-rich alkaline solutions with a low temperature gradient. Similar conditions were probably present when the Anahí quartz was deposited.

The Anahí Mine is the world's only commercial source not only of gemmy ametrine but also of natural citrine. It is also an important source of amethyst. Annual production at the beginning of the decade was about 3 million carats of cut stones plus 100 tonnes of low-grade amethyst for spheres, cabochons, beads, and tumbling and another 10 tonnes of specimens for the collector market. In 2008, 74 workers were active in five tunnels, and the annual production of gem rough was between 2,500 and 3,500 kilograms per 120 tonnes of quartz mined (Weldon, 2009).

Other Deposits

There are many other genetically similar amethyst occurrences in the La Gaiba area. Nearby Mina Pobre has also yielded ametrine, but of a much lower quality than that from the Anahí Mine. It has also produced amethyst-smoky quartz combinations. The Ayoreíta Mine, about 5 kilometers from Anahí, has no ametrine, but produces uniquely shaped amethyst crystals in vugs weighing up to several hundred kilograms. The individual crystals consist of a long prism terminated with two equant rhombohedra and often exceed 10 centimeters in length. Bicolor cut quartz from here is commonly violet and colorless rather than violet and yellow. Some of the low-quality amethyst is heated to produce bicolor, yellow-colorless, stones.

Ahlfeld and Muñoz Reyes (1955) describe a huge hydrothermal amethyst vein, 6 kilometers long and 4 to 6 meters wide, near the town of Santa Ana in Santa Cruz department, but the amethyst there is of low quality. Not far from Santa Ana, a vertical amethyst vein near the village of San Fermín was exploited a few years ago for rich purple amethyst.

***Right**: Slices of "ametrine" from the Anahí Mine The cross section on the right is 11 cm wide Private collection Photo Harold and Erica Van Pelt*

***Below**: Quartz, variety amethyst, group from the Anahí Mine 43 cm (~17 inches) wide! Rice Northwest Museum of Rocks and Minerals collection Jeff Scovil photo*

Bolivian Mineral Localities

by ***Alfredo Petrov, Bill Smith,*** *and* ***Carol Smith***

Place Name	Place Type	Lat. Long.	Province, Department	Known Mineral Species/ Notes
Aceromarca	mine		E slopes Taquesi batholith, Mururata dist., Sud Yungas prov., La Paz	Pyrite, pyrrhotite, scheelite, sphalerite, tourmaline, wolframite group minerals
Aguas Calientes	village		on the RR, Arque prov., Cochabamba	See La Salvadora 2
Aguila	mine		NW slopes of Illimani, La Paz	Stibnite
Aiquile	town	18°10'S 65°10'W	135 km SE of Cochabamba, Campero prov., Cochabamba	See Santa Marta
Alasca	mine		2 km from Pongo, Murillo prov., La Paz	Cassiterite, pyrrhotite, scheelite
Alcacoya	mine		San Vicente, Sud Chichas prov., Potosí	Stibnite
Aliada	mine		Colquechaca, Chayanta prov., Potosí	Teallite
Alto Chapare	district		~30 km W of Villa Tunari, Chapare prov., Cochabamba	Aegirine, aragonite, boracite, danburite, dolomite, dravite, ericaite, hematite, *magnesioriebeckite*, ("crocidolite"), magnesite, microcline, *povondraite*, pyrite; see Limbo
Amayapampa	town	18°29'S 66°26'W	40 km SE of Uncía, near Chayanta, Bustillo prov., Potosí	Gold, pyrite
Amigos	mine		Sud Lípez prov., Potosí	Stibnite
Amutara	mountain		30 km NW of Chicote, Inquisivi prov., La Paz	Quartz, wolframite group minerals
Anahí	mine		Near La Gaiba, Sandoval prov., Santa Cruz	Quartz ("ametrine"); see page 85
Ancohuma (west side)	peak (6,550 m)	15°50'S 68°35'W	Primary peak of Illampú massif, Larecaja prov., La Paz	Bismuth, bismuthinite, orthoclase, rutile
Ancoma	town	19°07'S 65°41'W	20 km NE of Sorata, Larecaja prov., La Paz	see Polo Sur, The Consolidated
Ancoraimes	town	15°54'S 68°58'W	113 km NW of La Paz, on shores of Lake Titicaca, Omasuyos prov., La Paz	Hübnerite, realgar; see Magariñas
Andacaba	mine, district		30 km SSE of Cerro Rico, in Kari-Kari range, Linares prov., Potosí	Arsenopyrite, boulangerite, canfieldite, cassiterite, cerussite, galena, *"potosíite"*, pyrite, siderite, sphalerite, wolframite group minerals; see Santa Barbara
Andamarca	city	18°46'S 67°31'W	105 km SW of Oruro, Sud Carangas prov., Oruro	see Calama
Andina	mine		Mururata dist., Sud Yungas prov., La Paz	Cassiterite, wolframite group minerals
Animas	mine, village	20°57'S 66°33'W	Animas-Chocaya subdistrict, Sud Chichas prov., Potosí	Apatite, *aramayoite*, argyrodite/canfieldite, bismuthinite, cassiterite, "franckeite", freibergite, galena, hocartite, jamesonite, marcasite, miargyrite, pyrite, quartz, siderite, stannite, stephanite, tetrahedrite, wurtzite, zinkenite; see Chocaya, Gran Chocaya
Animas-Chocaya	subdistrict		Atocha-Quechisla district, Sud Chichas prov., Potosí	see Chocaya, Oploca 1, Siete Suyos, Animas, Chocaya Grande
Antequera (aka El Salvador, Bolivar 2)	mine, village	18°28'S 66°50'W	13 km SSE of Poopó, Pazña dist., Poopó prov., Oruro	Cassiterite ("needle tin"), greenockite, pavonite, sphalerite, stannite, teallite, wurtzite, zinkenite; see Venus, Cerro Chunchu
Antofagasta	mine		Oruro dist., Cercado prov., Oruro	Stibnite with interstitial gold
Araca	town		Loayza prov., La Paz	see Viloco
Ascensión	mine		Condeauque dist., Cercado prov., Oruro	Azurite, barite, cerussite, hemimorphite, minium, scheelite, siderite, smithsonite
Asiento	mine		Mizqui prov., Cochabamba	See Quioma
Asiento de Araca	village		8 km WNW of Viloco, Loayza prov., La Paz	See Rosario 1
Asunta	mine		4 km NE of San Vicente, Sud Chichas prov., Potosí	Franckeite, greenockite, stannite
Atocha	town, district, river	20°56'S 66°14'W	80 km SE of Uyuni, Sud Chichas prov., Potosí	Minerals said to be from Atocha are actually from the Atocha-Quechisla district.

The most important localities are listed in red; type species for listed localities are shown in *italics*; "?" indicates a questionable designation.

***Above**: Danburite; 3.6 cm tall; Alto Chapare; John White collection (ex Alfredo Petrov); Jeff Scovil photo*

***Upper middle**: Boracite; the crystal is 1.2 cm wide; Alto Chapare; Alfredo Petrov collection; Jeff Scovil photo*

***Lower middle**: Aramayoite; 3.5 cm tall; Animas Mine, Sud Chichas province, Potosí; Terry Wallace collection; Jesse La Plante photo*

***Upper right**: Wurtzite and galena; 11.5 cm tall; Animas Mine; Rock Currier collection; Jeff Scovil photo*

***Lower right**: Freibergite and pyrite; 16.2 cm wide; Animas Mine; Rock Currier collection, Jeff Scovil photo*

***Below**: Apatite with siderite and pyrite; xl 1.7 cm wide; Colorado vein, Animas Mine; Dave Bunk specimen; Jesse La Plante photo*

Place Name	Place Type	Lat. Long.	Province, Department	Known Mineral Species/ Notes
Aullagas	ghost town		on Colquechaca Mountain, Chayanta prov., Potosí	see Colquechaca
Avicaya	mine	18°50'S 67°00'W	10 km NNE of Pazña, Pazña dist., Poopó prov., Oruro	Cassiterite, jarosite, stannite, vivianite
Ayoreíta	mine		La Gaiba dist., Sandoval prov., Santa Cruz	Amethyst, hollandite, smoky quartz
Berenguela	mine		Cerro Grande dist., Arque prov., Cochabamba	Cassiterite, pyrite, pyrrhotite, silver sulfosalts
Berenguela	mine, town, district	17°17'S 69°14'W	22 km SE of Santiago de Machaca, Pando prov., La Paz	Apatite, greenockite, quartz, siderite, sphalerite; see Jocullani, San Luis 1
Bolivar 1	mine	21°48'S 66°27'W	S flank Cerro Bonete, E. Sud Lípez prov., Potosí	Aikinite, ankerite, benjaminite, berryite, bismuthinite, chalcopyrite, krupkaite, *pavonite*, pyrite, siderite
Bolivar 2	mine		Pazña dist, Poopó prov., Oruro	see Antequera, El Salvador
Bolivia	mine		5 km east of Poopó, Poopó prov., Oruro	Jamesonite

Place Name	Place Type	Lat. Long.	Province, Department	Known Mineral Species/ Notes
Bolsa Negra (aka Nevada)	mine		Yanacachi subdistrict, Sud Yungas prov., La Paz	Ferberite, scheelite, wolframite
Bombori	village		Chayanta prov., Potosí	Amethyst, barite, calcite
Cacachaca	village		ENE of Challapata, Avaroa prov., Oruro	Chalcostibite; see Rosita, Terremoto
Cachilaguna	dry lake	21°44'S 67°57'W	On the border of Nor Lípez and Sud Lípez provinces, Potosí	Halite, natron, thermonatrite, trona, ulexite
Calacalani (aka San Antonio de Calacalani)	mine		Cercado prov., Oruro	see San Antonio 1
Calama	town	18°44'S 67°40'W	13 km WNW of Andamarca, Sud Carangas prov., Oruro	Bismuth, gold, gypsum, silver
Caluyo	town	16°49'S 68°06'W	48 km SSE of Inquisivi, Inquisivi prov., La Paz	Stibnite; see Pacopampa, Titina
Candelaria	mine		7 km S of Poopó, Poopó prov., Oruro	Siderite, teallite, "montesite", wurtzite
Canutillos	mine		Colavi, Saavedra prov., Potosí	Barite, bournonite, cassiterite, ludlamite, opal, pyrite, siderite, vivianite
Capa Circa (aka Capacirca)	mine		24 km SE of Uncía, Bustillo prov., Potosí	Gold, stibnite
Carabuco	mine		Choquetanga dist., Inquisivi prov., La Paz	Scheelite (pink to orange, with tourmaline inclusions)
Caracoles	mine, village	16°59'S 67°24'W	15 km W of Inquisivi, Inquisivi prov., La Paz	Ankerite, apatite, bismuth, bismuthinite, cassiterite, chlorite, pyrite, tourmaline
Caracollo	town	17°39'S 67°10'W	39 km NW of Oruro, Cercado prov., Oruro	see Tarumita
Caracota	mine		25 km ESE of Yura, Quijarro prov., Potosí	Gold, stibnite
Carangas	mine	18°56'S 68°38'W	29 km ESE of Sabaya, Mejillones prov., Oruro	Chlorargyrite, rhodochrosite, silver, silver sulfosalts
Carguaicollo (aka Carguaycollo, Carhuaycollo)	town	19°31'S 66°40'W	80 km S of Huanchaca, Quijarro prov., Potosí	Anglesite, cassiterite ("needle tin"), franckeite, greenockite, pyrite, quartz, sphalerite, stibiconite, teallite, wurtzite; see Cerillos
Carmen Pampa	village		15 km S of Coroico, Nor Yungas prov., La Paz	Siderite, tetrahedrite
Carmen (aka Carmen-Aurora)	mine		La Unión district, Los Andes prov., La Paz	Bindheimite, bismuth (source of Vilaque River nuggets), bismuthinite, cassiterite, jamesonite, pyrite
Cascabel	mine		Quiabaya, Larecaja prov., La Paz	Anglesite, boulangerite, galena, jamesonite
Cataricahua	mine		E slopes of Mt. Pozoconi; near Huanuni Mine, Dalence prov., Oruro	Cassiterite, goethite, scorodite
Catavi	mill		3 km NE of Llallagua, Bustillo prov., Potosí	Minerals ascribed to Catavi came from the Siglo XX Mine (Llallagua)

***Above**: Bismuth and bismuthinite; 10.2 cm tall*
Carmen Mine, La Unión district, Los Andes province, La Paz
Bill and Carol Smith collection
Jeff Scovil photo

***Above**: Quartz; 9.3 cm wide*
Chicote Grande, Inquisivi province, La Paz
Francis Benjamin collection; Jeff Scovil photo

***Left**: Ferberite; 15.5 cm tall; Chambillaya, Inquisivi prov., La Paz; Canadian Museum of Nature collection; Michael Bainbridge photo*

Place Name	Place Type	Lat. Long.	Province, Department	Known Mineral Species/ Notes
Cebadillas	mine		near Chuquiuta, Bustillo prov., Potosí	Gold, stibnite
Cerillos	small mountain		near Carguaicollo, Quijarro prov., Potosí	Teallite
Cerro Bonete	mountain		20 km N of Esmoraca, Sud Lípez prov., Potosí	see Rosario 2, Bolivar 1
Cerro Chunchu	mountain		near Antequera, Poopó prov., Oruro	Cassiterite (some Fe-rich "varlamoffite"), hydrocerussite
Cerro Colorado	mountain	21°37'S 65°16'W	Cotagaita, Nor Chichas prov., Potosí	Cinnabar
Cerro Grande	mine, district		60 km SW of Cochabamba, Arque prov., Cochabamba	Apatite, cassiterite, siderite; see Berenguela
Cerro Malmisa	mountain		SE of Maragua, Chayanta prov., Potosí	Antimony, ferberite, molybdenite
Cerro Negro 1	mountain		15 km SE of Illimanì, Sud Yungas prov., La Paz	Pyrophyllite, wolframite group minerals
Cerro Negro 2	mountain		E of Vichacla, Nor Chichas prov., Potosí	Annabergite, gersdorffite, löllingite
Cerro Pozoconi	mine, mountain		Huanuni, Dalence prov., Oruro	See Pozokoni, also written Posokoni
Cerro Rico (aka Cerro de Potosí)	mountain		Potosí, Bustillo prov., Potosí	Acanthite, alunite, andorite, apatite,arsenic, arsenopyrite, arsenosulvanite, barite, *berndtite*, bismuthinite, bournonite, calcite, cassiterite, chalcopyrite, chlorargyrite, diaphorite, franckeite, freibergite, galena, goethite (iridescent), goslarite, halotrichite, hematite, jamesonite, jarosite, matildite, melanterite, miargyrite, nacrite, *ottemannite*, phosphophyllite, pickeringite, pyrargyrite, pyrite, pyrostilpnite, quartz, siderite, sideronatrite, siderotil, silver, smithsonite, sphalerite, stannite, stephanite, stibnite, tetrahedrite, valentinite, voltaite, wavellite, wolframite, wurtzite; see page 36
Cerro San Pedro	hill		Oruro, Cercado prov., Oruro	Large orthoclase phenocrysts, often replaced by quartz, chlorite, siderite, etc.
Cerro Sapo	mountain		18 km N of Independencia, Ayopaya prov., Cochabamba	Amesite-2H, analcime, ankerite, barite, chalcopyrite, dawsonite, galena, krasnovite, natrolite, sodalite; see page 80
Cerro Ubina	mountain		near Ubina, Quijarro prov., Potosí	Bismuthinite, pyrite, tourmaline
Cerro Vilacota	mountain		between Acacio and San Pedro de Buena Vista, Charcas prov., Potosí	Barite, galena
Chacaltaya	mountain (4,640 m), mine	16°20'S 68°08'W	40 km N of La Paz, Murillo prov., La Paz	Arsenopyrite, cassiterite, fluorite, muscovite, nacrite, ullmanite (As-rich "korynite"); see San Carlos 2
Chacarilla	town	17°50'S 68°00'W	30 km NW of San Pedro de Curahuara, Villaroel prov., La Paz	Copper, chalcocite, malachite
Chajolpaya	village		15 km E of Mt. Illampu, Larecaja prov., La Paz	Cassiterite
Challahuilque	village		just N of San Cristobal, Nor Lípez prov., Potosì	Cerussite, chlorargyrite
Challana	river, village		Larecaja prov., La Paz	Columbite, heterosite, rutile, spodumene, triphyllite, and triplite are found in the vicinity; see Fabulosa
Challapata	town	18°54'S 66°47'W	113 km SSE of Oruro, Avaroa prov., Oruro	Chalcostibite, teallite, wurtzite; see Terremoto, Vicuñita
Challviri	town		Ibañez prov., Potosí	Stibnite
Chambillaya	mine		near Quime, Inquisivi prov., La Paz	Apatite, arsenopyrite, quartz twins, wolframite
Chayanta	town	18°27'S 66°30'W	13 km E of Uncía, Bustillo prov., Potosí	Apatite, cassiterite, silver; see Capa Circa, Pucro, Vila Apacheta
Chicote Chico	mountain		1 km S of Chicote Grande, Ayopaya prov., Cochabamba	Wolframite; see Tamiñani
Chicote Grande	mountain, mine		8 km NW of Kami, Inquisivi prov., La Paz	Arsenopyrite, ferberite, muscovite, pyrite, quartz (Japan-law twins), scorodite, siderite, tungstite; see Lapiani
Chijmuni	village	17°20'S 67°50'W	10 km WNW of Sica Sica, Aroma prov., La Paz	see Lili, Laurani, and Lomitas, all of which are near Chijmuni

Above: *Sulfur; 8.7 cm wide*
El Desierto Mine, Daniel Campos province, Potosí
Klaus Hedegaard collection; Jeff Scovil photo

Right: *Siderite and pyrite on quartz; 9 cm tall*
Chojñacota Mine, Loayza province, La Paz
Dave Bunk Minerals specimen; Jesse La Plante photo

Above: *Gold nugget*
4.2 cm tall, 3.1 ounces
Chuquiaguillo River,
Murillo prov., La Paz
The Kilian collection
Jeff Scovil photo

Place Name	Place Type	Lat. Long.	Province, Department	Known Mineral Species/ Notes
Chilcani	mine		near Lequepalca, Cercado prov., Oruro	Bindheimite, chalcostibite, gold in stibnite
Chilcobija	mine		Potosí, Bustillo prov., Potosí	Stibnite, chalcostibite
Chocaya	mine, town		Animas-Chocaya subdistrict, Sud Chichas prov., Potosí	Apatite, aramayoite, argyrodite, arsenopyrite, barite, bismuth, cassiterite, “elaterite,” franckeite, *hocartite*, jarosite, marcasite, pyrite, quartz, *ramdohrite*, stannite, tetrahedrite, wavellite, wurtzite; see Animas, Gran Chocaya, Siete Suyos, Oploca 1
Chocaya Grande	mine		Animas-Chocaya subdistrict, Sud Chichas prov., Potosí	Pyrargyrite
Chojlla	mine, town		16 km W of Caripuyo, Yanacachi subdistrict, Alonso de Ibanez prov., Potosí	Brown to pink cassiterite, feldspar, ferberite, muscovite, quartz, tungstite, wolframite; see Socavón Carmen
Chojñacota	mine		Loayza prov., La Paz	Arsenopyrite, cassiterite, chalcopyrite, pyrite, quartz, siderite, sphalerite
Choquecamata	town	16°55’S 66°37’W	13 km NE of Cocapata, Ayopaya prov., Cochabamba	Gold
Choquetanga	mine, mountain, district		100 km SE of La Paz, 20 km NW of Quime, Inquisivi prov., La Paz	Scheelite; see Carabuco
Chorolque	mine, mountain, district	20°55’S 66°02’W	35 km E of Atocha, Atocha-Quechisla dist., Sud Chichas prov., Potosí	Bismuth, bismuthinite, bournonite, cassiterite (some after apatite, pyrite, and bismuthinite), daubréeite, ferberite, goethite, gold, jamesonite, pyrite, scorodite, tourmaline, tungstite; see Espíritu, La Reforma, San Miguel 2, San Carlos 1
Chulchucani	deposit		25 km NE of Potosí, near Huari Huari, Frías prov., Potosí	see San Luis 2
Chuma	town	15°30’S 68°56’W	145 km NNW of La Paz, Muñecas prov., La Paz	see Hualpani
Chuquiaguillo	river		5 km N of La Paz, Murillo prov., La Paz	Gold nuggets to 7 kg
Chuquiuta	town	18°33’S 65°26’W	12 km SW of Amayapampa, Bustillo prov., Potosí	see Cebadillas
Churiña (aka San Pablo de Churiña)	mine		Macha, Chayanta prov., Potosí	Siderite, stibnite
Churquini	mine		Pampa Grande district, Sud Chichas prov., Potosí	Galena, hydrozincite, linarite, sphalerite, stibnite
Chusi	village		Tipuani, Larecaja prov., La Paz	Gold
Cobrizos	mine	20°60’S 69°13’W	San Cristobal, Nor Lípez prov., Potosí	Azurite, chalcocite, copper, gypsum, malachite, silver
Coipasa	town	19°17’S 68°17’W	20 km S of Sabaya, Atahuallpa prov., Oruro	Halite, ulexite

Place Name	Place Type	Lat. Long.	Province, Department	Known Mineral Species/ Notes
Colavi (aka Machacamarca)	mine, town, district	19°21'S 65°33'W	36 km NE of Potosí, Saavedra prov., Potosí	*Bismite*, goethite after siderite, opal, pyrite, siderite, valentinite; see Canutillos, Mosojllacta, Machacamarca
Colquechaca	city, mountain	18°42'S 66°00'W	105 km NNW of Potosí, Chayanta prov., Potosí	Acanthite, argentopyrite, argyrodite, arsenopyrite, barite, bismuth, bournonite, calcite, *canfieldite*, cassiterite, chalcostibite, cylindrite, dyscrasite, "elaterite", franckeite, galena, hocartite, jamesonite, löllingite, marcasite, miargyrite, polybasite, proustite, pyrargyrite, pyrite, pyromorphite, quartz, siderite, silver (some after pyrargyrite and stephanite), sphalerite, stephanite, vivianite, wurtzite; see also Aliada, Nueva Virginia, Aullagas; page 63
Colquiri	mine, town	17°25'S 67°08'W	78 km N of Oruro, Inquisivi prov., La Paz	Cassiterite, *colquiriite*, creedite, fluorite, gearksutite, gypsum, hocartite, madocite, pyrite, ralstonite, sphalerite; see San Antonio 1
Comanche	town	17°01'S 68°27'W	10 km SW of Collana, Pacajes prov., La Paz	Chalcocite and malachite; see Miriquiri
Concordia (aka Santo Domingo)	mine		ENE slopes of Nevada Cunocallo, Cordillera Santa Vera Cruz, Inquisivi prov., La Paz	Cassiterite, tourmaline; elevation to 5,060 m
Condeauque (aka Conde Auque)	district		NE of Oruro, Cercado prov., Oruro	Dolomite, fluorite, scheelite, wolframite; see Ascención, Santa Isabel
Consolidated, The	mine		20 km NE of Sorata, Ancoma, Larecaja prov., La Paz	Arsenopyrite, cassiterite, chalcopyrite, muscovite, pyrrhotite, scheelite, tourmaline, tungstite
Constancia	mine		Tasna, Sud Chichas prov., Potosí	Daubréeite
Coricoya	mine		near Molínero, Mizque prov., Cochabamba	Bornite, brochantite
Coriviri	mine		between Poopó and Machacamarca, Poopó prov., Oruro	Bismuth, bismuthinite, gold, wolframite
Corocoro	mine, district	17°12'S 68°29W	89 km SSW of La Paz, Pacajes prov., La Paz	Algodonite, aragonite, azurite, celestine, chalcophyllite, copper, copper after aragonite, cuprite, domeykite, fluorite, gypsum, jamesonite, malachite, silver, tenorite, tetrahedrite; see Marcocoya, Pucara, San Agustín 1, San Francisco 2, Remedios, page 76
Coroico	city	16°11'S 67°44'W	55 km NE of La Paz, Nor Yungas prov., La Paz	see Carmen Pampa
Cotagaita	town, river	20°50'S 65°37'W	65 km N of Tupiza, Nor Chichas prov., Potosí	see Churquini, Pampa Grande, Cerro Colorado, Tapi
Cristalmayu	creek		between Filadelfia and San Francisco Mines, Alto Chapare dist., Chapare prov., Cochabamba	see Alto Chapare
Cuprita	mine		Turco, Sajama prov., Oruro	Chalcedony, copper
David Mosiah	mine		near El Molino on Mt. Huañaquino, Frías prov., Potosí	Barite, hematite, apatite-(CaOH), ilmenorutile, magnetite, schorl, rutile after ilmenite; pg. 48
Don Diego	village		20 km NE of Potosí, border of Frías and Saavedra provinces, Potosí	see Tricolore
Dos Amigos	small mine		E slope of Cord. Santa Vera Cruz, Cercado prov., Oruro	Antimony (some after stibnite)
El Barco	mine		Yanacachi, Sud Yungas prov., La Paz	Cassiterite
El Desierto	mine		8 km S of San Pablo de Napa, Daniel Campos prov., Potosí	Alum-K, sulfur
El Dragón	mine		E of Porco near Sala Kuchu, Quijarro prov., Potosí	Ahlfeldite, allophane, alunite, anglesite, barite, basaluminite, berzelianite, chalcomenite, chalcopyrite, clausthalite, clinochalcomenite, dolomite, eskebornite, esperite, kerstenite, klockmannite, krutaite, lepidocrocite, linarite, malachite, molybdomenite, olsacherite, penroseite, reevesite, schmiederite, selenium, tiemannite, trogtalite, umangite

Place Name	Place Type	Lat. Long.	Province, Department	Known Mineral Species/ Notes
El Molino	village		15 km NW of Potosí, Frías prov., Potosí	see David Mosiah
El Paraiso	mine		San Ramón, Ñuflo de Chavez prov., Santa Cruz	Ferrocolumbite
El Salvador	mine		Pazña dist, Poopó prov., Oruro	see Antequera
Empexa	salt flats	20°20'S 68°29'W	Campos prov., Potosí	Ulexite; see Sijsihua
Esmoraca	town	21°42'S 66°15'W	40 km SE of San Pablo, Sud Chichas prov., Potosí	see Pueblo Viejo, Sucre, Rosario 2
Espíritu	mine		Chorolque, Sud Chichas prov., Potosí	Bismutite
Fabulosa	mine		10 km upstream of Challani Village, E flank Cord. Real, Larecaja prov., La Paz	Almandine, andalusite, arsenopyrite, cassiterite, chalcopyrite, cordierite (iolite), lazulite, molybdenite, muscovite, pyrite, pyrrhotite, quartz, sphalerite, stannite, triphylite, triplite, vivianite
Farellones	mine		Tasna, Sud Chichas prov., Potosí	Bismuthinite, ferritungstite, tungstite after ferberite
Fenix	mine		Cercado prov., Oruro	see Juliana
Filadelfia	mine		Alto Chapare dist., Chapare prov., Cochabamba	Magnesioriebeckite, hematite, danburite, dolomite; see Alto Chapare
Gran Chocaya	mine		Animas-Chocaya subdistrict, Sud Chichas prov., Potosí	Pyrargyrite
Gruta de San Pedro	cavern		San Pedro, Larecaja prov., La Paz	Remarkable 1 km plus long karst cavern, eroded from gypsum rock.
Guadaloupe	mining company			The company that operated Tatasí from 1890 to 1902
Gutierrez	mine		Porco, Quijarro prov., Potosí	Cerussite
Hiaco	mine		20 km NE of Colquechaca, Chayanta prov., Potosí	Barite, cerussite, galena, zincian argentite; see Pacajake
Huallatani	mine		Huayna Potosí, Los Andes prov., La Paz	Bismuth, cassiterite
Hualpani	mine		Chuma, Muñecas prov., La Paz	Anglesite, galena, sphalerite
Huanchaca (aka Pulacayo Mine)	mine, town	20°20'S 66°39'W	30 km NE of Uyuni, Near Pulacayo, Quijaro prov., Potosí	see Pulacayo
Huañaquino	mountain		Frías prov., Potosí	see David Mosiah, San Calixto; page 48
Huanuni	mine, town	18°17'S 66°51'W	45 km SSE of Oruro, Dalence prov., Oruro	*Aheylite*, apatite, beidellite, bismuthinite, bournonite, cassiterite, chalcanthite, cronstedtite, ferberite, fibroferrite, fluorite, franckeite, halloysite, herzenbergite, hisingerite, hydrozincite, jamesonite, jarosite, kaolinite, linarite (?), ludlamite, marcasite, *nikischerite*, opal, owyheeite, plagionite, pyrite, pyrrhotite, semseyite, siderite, sphalerite, stannite, tourmaline, valentinite, variscite, vivianite, wavellite, wurtzite (?); see María Francisca, Miraflores, Perez, Cerro Pozokoni, Porvenir 1, Quimsacoya; page 18
Huari	town	18°46'S 65°41'W	121 km SSE of Oruro, Avaroa prov., Oruro	see María Teresa
Huari Huari	town, range	19°25'S 65°38'W	25 km NE of Potosí, Frías prov., Potosí	Pyromorphite; see Orco Cuchillo
Huayna Potosí (aka Huayna, Huaina)	village	16°16'S 68°11'W	Los Andes prov., La Paz	see Carmen, Huallatani
Huayoco	gully, village		25 km SSE of Macha, Condor Iquiña Plateau, Chayanta prov., Potosí	"wood tin"
Hucumarini	mine		Sorata, Larecaja prov., La Paz	Bismuthinite
Ichucolla	mine		Pazña dist., Poopó prov., Oruro	old name for Monserrat Mine
Illimani	mountain (6,457 m)	16°39'S 67°48'W	45 km ESE of La Paz, Murillo and Sud Yungas provinces, La Paz	see Urania, Palca
Inquisivi	city	16°41'S 67°10'W	115 km ESE of La Paz, Inquisivi prov., La Paz	see Caluyo
Iroco	mine		Oruro, Cercado prov., Oruro	Gold, goethite, scorodite

***Left**: Quartz and scorodite; 9.3 cm tall; Kami Mine, Ayopaya province, Cochabamba; Rock Currier collection; Jeff Scovil photo*

***Middle**: Scorodite; 11 cm tall; Kami district; Marty Zinn collection; Jeff Scovil photo*

***Above**: Bournonite cluster; 2.6 cm tall; Colavi (Machacamarca) district, Saavedra province, Potosí; Max and Jon Sigerman collection; Budd photo*

Place Name	Place Type	Lat. Long.	Province, Department	Known Mineral Species/ Notes
Iscaisca	mountain		28 km N of Tupiza, Nor Chichas prov., Potosí	Semseyite; see Kentiyok
Itos	mine		Oruro, Cercado prov., Oruro	Acanthite, arsenopyrite, augelite, andorite, cassiterite, chalcostibite, dolomite, franckeite, freibergite, heteromorphite, jamesonite, kaolinite, metastibnite, miargyrite, natroalunite, pyrite, silver, stannite, stibnite, tetrahedrite, zinkenite
Jachahuance	mine		Choquetanga, Inquisivi prov., La Paz	Quartz
Japo	mine		Dalence prov., Oruro	see Negro Pabellón
Jocullani	mine		Berenguela, Pando prov., La Paz	Sphalerite
Juliana (aka Fenix)	mine		on Cochabamba border, Cercado prov., Oruro	Ankerite, hydrotungstite, scheelite, tungstite; see La Bomba
Kami	mountain, town, mine	17°22'S 66°58'W	7 km SSE of Chicote Grande, Ayopaya prov., Cochabamba	Apatite, arsenopyrite, cassiterite, covellite, ferberite, jarosite, jeromite, lazulite, pyrite, pyrrhotite, quartz (Japan-law twins), scorodite, siderite, tungstite, tourmaline, wolframite, wurtzite
Kari-Kari	mountain range	19°40'S 65°40'W	E and SE of city of Potosí, mostly in Frías prov., Potosí	Almandine, epidote, hopeite; see Andacaba
Kentiyok	mine		2 km N of Mt. Iscaisca, Nor Chichas prov., Potosí	Galena
Kesñiri	mine		Between Toro Toro and San Pedro de Buena Vista, Charcas prov., Potosí	Barite, galena, pyromorphite
La Bella	mine		40 km E of San Ramon, San Ramón dist., Ñuflo de Chavez prov., Sant Cruz	Beryl, columbite, fergusonite, fluorite, monazite, topaz
La Bomba	mine		2 km NW of Juliana Mine, Cercado prov., Oruro	Fluorite, scheelite
La Colorada	mine		Oruro, Cercado prov., Oruro	Bournonite, cassiterite, franckeite
La Escosasa	mine		Patacamaya, Aroma prov., La Paz	Galena
La Gaiba	village, lake, district	17°45'S 57°43'W	30 km N of Puerto Suarez, on the border w/Brazil, Sandoval prov., Santa Cruz	see Anahí, Ayoreíta
La Joya	mountain	17°46' S 67°29'W	48 km NW of Oruro, Cercado prov., Oruro	Gold, muscovite
La Ponderosa	mine		Turco, Sajama prov., Oruro	Chalcedony, copper
La Reforma	mine		Chorolque, Sud Chichas prov., Potosí	Bismuth, bismuthinite

Place Name	Place Type	Lat. Long.	Province, Department	Known Mineral Species/ Notes
La Salvadora 1	mine		Oruro, Cercado prov., Oruro	La Salvadora is the name of the cooperative operating the Socavón. See Socavón
La Salvadora 2	prospect		Aguas Calientes, Arque prov., Cochabamba	Barite, galena, stibnite
La Salvadora 3	mine		~40 km E of La Paz, E flank of Taquesi batholith, Sud Yungas prov., La Paz	Apatite, cuprotungstite, scheelite, tourmaline, wolframite
La Sorpresa	mine		10 km S of Tapacarí, Tapacarí prov., Cochabamba	Annabergite, gersdorffite, nickeline
La Unión	district		W of Milluni, Los Andes prov., La Paz	see Carmen-Aurora
La Verde	mine		5 km NW of San Agustín 2, 35 km E of San Ramón, Ñuflo de Chavez prov., Santa Cruz	Columbite, topaz, green muscovite
Laguna Cachi	dry lake		On the border of Nor Lípez and Sud Lípez provinces, Potosí	see Cachilaguna
Lapiani	mine		Chicote Grande, Inquisivi prov., La Paz	Pyrite, siderite
Laurani	mine	17°23'S 67°47'W	5 km WSW of Sica Sica, Aroma prov., La Paz	*Braithwaiteite*, enargite, *lammerite*, lavendulan, luzonite
Lequepalca	village	17°38'S 66°55'W	30 km E of Caracollo, Cercado prov., Oruro	Ferberite, hydrozincite, linnaeite, stibiconite, stibnite; see San Miguel 3
Lili	quarry		Chijmuni, Aroma prov., La Paz	Austinite ("brickerite"), mimetite, pitticite, "travertine"
Limbo	village (uninhabited)		Alto Chapare dist., Chapare prov., Cochabamba	Actinolite, anhydrite, aragonite, augite, magnesioriebeckite, hydromagnesite, magnesite, pyrite, talc
Lípez Huayco	mine		Ocurí, Chayanta prov., Potosí	Cassiterite (some after teallite), herzenbergite, ("montesite"), teallite
Llallagua	town	18°25'S 66°38'W	5 km NNW of Uncía, Bustillo prov., Potosí	Allophane, alunite, apatite-(CaF), arsenic, arsenopyrite, augelite, bismite, bismoclite, bismuth, bismuthinite, bismutite, brochantite, cassiterite, chalcanthite, chalcocite, childrenite, chrysocolla, copper, cordierite, crandallite, creedite, cronstedtite, cylindrite, diadochite, evansite, ferberite, florencite, franckeite, goethite, greenockite, hagendorfite, halotrichite, hematite, hisingerite, hübnerite, jamesonite, jarosite, *jeanbandyite*, kaolinite, marcasite, melanterite, *metavauxite*, miargyrite, monazite, muscovite (after cordierite), natantite, orthoclase, *paravauxite*, pickeringite, plumbogummite, pyrite, pyromorphite, pyrrhotite, quartz (Japan-law twins), rhodochrosite, scheelite, sericite after orthoclase, siderite, *sigloite*, sphalerite ("gumucionite"), stannite, stibnite, teallite, tetrahedrite, tourmaline, variscite, *vauxite*, vivianite, wavellite, wickmanite, wurtzite, xenotime, zinkenite. Almost all of the Llallagua minerals came from the Siglo XX Mine.
Llavisa	village		16 km SW of Ocurí, Chayanta prov., Potosí	Adularia, cassiterite ("wood tin"), gersdorffite, topaz
Lomitas	"Little Hill", RR station	17°24'S 67°50'W	10 km SSW of Sica Sica, Aroma prov., La Paz	Austinite ("brickerite"), chalcedony, mimetite, pitticite; see nearby Lili and Laurani
Luribay	town	17°14'S 67°44'W	90 km SE of La Paz, Loayza prov., La Paz	See Viloco, Yaco
Macha	town	18°49'S 66°05'W	16 km SSW of Colquechaca, Chayanta prov., Potosí	See Milluri, Churiña, Huayoco, Santiaguillo; these "wood tin" localities are often given on labels as "near Macha"
Machacamarca (NOT Litoral prov., Oruro)	ghost town, mine	19°24'S 65°33'W	30 km NE of Potosí, Colavi district, Saavedra prov., Potosí	Augelite, chlorargyrite, bindheimite, bournonite, diaphorite, freibergite, jamesonite, pyrite, pyromorphite,quartz, scorzalite, siderite, tetrahedrite after bournonite; see Víboras

Place Name	Place Type	Lat. Long.	Province, Department	Known Mineral Species/ Notes
Machacamarca (aka Colavi)	district		Saavedra prov., Potosí	see Colavi
Magariñas	mine		Ancoraimes, Larecaja prov., La Paz	Ferberite, geocronite
Maragua	town	18°56'S 65°48'W	8 km S of Ocuri, Sorata district, Chayanta prov., Potosí	Cassiterite; see Santiaguillo, Vetillas, Cerro Malmisa
Marcamarcani (formerly San Baldomero)	abandoned mine		10 km S of Sorata, Larecaja prov., La Paz	Arsenopyrite, tellurian bismite, bismuth, bismuthinite, bismutoferrite, erythrite, joseite, löllingite, skutterudite ("smaltite")
Marcocoya	mine		Corocoro district, Pacajes prov., La Paz	Copper after aragonite
Margareta	mine		Quime, Inquisivi prov., La Paz	Vivianite, wolframite group minerals
María Francisca	mine		Huanuni, Dalence prov., Oruro	Cylindrite, (?) *franckeite*, jamesonite, stannite
María Teresa	mine		Huari, Avaroa prov., Oruro	Berndtite (?), *herzenbergite*, ottemannite (?), stannite
Mariana	mine		Tasna, Sud Chichas prov., Potosí	Jamesonite
Martha	mine		Pongo, Murillo prov., La Paz	Zinnwaldite
Matilde	mine		12 km NE of Ancoraimes, Omasuyos prov., La Paz	Cerussite, galena, siderite, sphalerite
Mercedes	mine		Alto Chapare, Chapare prov., Cochabamba	Magnesioriebeckite
Mesa de Plata	mine		San Antonio de Lípez, Sud Lípez prov., Potosí	Acanthite, andorite, bromargyrite, pyrargyrite, silver, tenorite
Millipaya	village		10 km SE of Sorata, Larecaja prov., La Paz	Dravite, hedleyite, jamesonite, molybdenite, molybdite, monazite, pyrite, quartz, schorl, tungstite, wolframite; see Marcamarcani, Hucumarini
Milluni	village		20 km N of La Paz, Los Andes prov., La Paz	Cassiterite; specimens ascribed to Milluni can be from either Los Andes or Murillo province
Milluri	mine		Macha, Chayanta prov., Potosí	Cervantite
Miraflores	neighboorhood		neighborhood in Huanuni, Dalence prov., Oruro	Probably not a true mineral locality. The single find of "blue lace" vauxites attributed to Miraflores have exact counterparts from the Siglo XX Mine in Llallagua, which is the likely source of the "Miraflores" (Huanuni Mine) specimens as well. With the excption of that find, no vauxite or associated species have ever been reported from Huanuni.
Miriquiri	mountain	16°58'S 68°25'W	near Comanche, Pacajes prov., La Paz	Apatite, hematite, magnetite
Mizque	city	17°56'S 65°19'W	115 km SE of Cochabamba, Mizque prov., Cochabamba	see Quioma
Mochará	village		24 km NE of Tupiza, Sud Chichas prov., Potosí	see Pirpintuyok
Mocoya (aka Tanapaca)	mine		3 km NW of Viloco, Loayza prov., La Paz	see Tanapaca
Mojo	village, district	21°54'S 65°33'W	30 km N of Villazón, Omiste prov., Potosí	see Volcano
Molínero	village	18°15'S 65°24'W	50 km S of Mizque, Mizque prov., Cochabamba	see Coricoya
Molino Mine	mine		Frías prov., Potosí	incorrect name for David Mosiah Prospect
Monserrat (formerly Ichucolla)	mine		Pazña dist., Poopó prov., Oruro	Brackebuschite, cassiterite (needle tin), cylindrite, franckeite, galena, "hochschildite," pyrite, stannite, *teallite (?)*, vivianite, wurtzite, zinkenite; see Santa Rosa
Moquegua	mine		Ubina, Quijarro prov., Potosí	Pyrite, tourmaline
Morococala	mine	18°10'S 66°44'W	40 km ESE of Oruro, Santa Fe dist., Dalence prov., Oruro	Cassiterite, hisingerite, phosphophyllite, pyrite, siderite, vivianite; see page 32
Moscari	town	18°19'S 66°02'W	11 km SSW of San Pedro, Charcas prov., Potosí	Ferberite, stibnite, wurtzite
Mosiah	mine		Frías prov., Potosí	see David Mosiah
Mosojllacta	mine		Colavi, Saavedra prov., Potosí	Bismuth, bismuthinite, goethite, opal, pyrite, siderite

Place Name	Place Type	Lat. Long.	Province, Department	Known Mineral Species/ Notes
Mururata	village	16°07'S 67°45'W	20 km NW of Coroico, Nor Yungas prov., La Paz	see Bolsa Negra
Mururata (includes the village of Mururata)	mountain, district	16°30'S 67°50'W	31 km ESE of La Paz, Nor and Sud Yungas provinces, La Paz	Molybdenite, scheelite, tungstite, wolframite; see Andina, Bolsa Negra, Aceromarca, Chojlla
Mutún	mountain		30 km SSW of Puerto Suarez, Busch prov., Santa Cruz	Cryptomelane, hausmanite, hematite, magnetite (30 billion tons of iron ore), pyrolusite
Napa (aka San Pablo de Napa)	village (abandoned)		Campos prov., Potosí	Alum-(K), sulfur, ulexite
Negra	mine	21°02'S 68°18'W	8 km NNE of Cosca, Chile, Nor Lípez prov., Potosí	Arseniosiderite, pyrolusite
Negro Pabellón (aka Japo)	mine	15°00'S 66°58'W	30 km SE of Oruro, Santa Fé dist., Dalence prov., Oruro	Cassiterite, hisingerite, pyrrhotite, siderite, vivianite
Nueva Venecia	ranch		Ñuflo de Chavez prov., Santa Cruz	Columbite, ilmenite (to 3.5 cm)
Nueva Virginia	mine		Colquechaca,Chayanta prov., Potosí	Cylindrite
Obrajes	hot springs		20 km NE of Oruro, Cercado prov., Oruro	Jarosite, thermonatrite
Ocurí	town	18°50'S 65°50'W	Chayanta prov., Potosí	Cassiterite, pyrargyrite, stibiconite; see Lípez Huayco
Oploca	mine		8 km NE of Chocaya, Animas-Chocaya subdistrict, Sud Chichas prov., Potosí	Argyrodite/canfieldite, cassiterite, hocartite
Oploca	village	21°19'S 65°46'W	16 km NNW of Tupiza, Sud Chichas prov., Potosí	Marcasite on pyrite, stibnite
Orco Cuchillo	mine		Huari Huari, Frías prov., Potosí	Marcasite, pyromorphite
Orlandini (mine owner's name)	mine		Ubina, Quijarro prov., Potosí	Cassiterite, pyrite, tourmaline
Oruro	town, department	17°59'S 67°08'W	Cercado prov., Oruro	Alunite, andorite, arsenopyrite, augelite, barite, boulangerite, bournonite, cassiterite, cerussite, chalcopyrite, chalcostibite, cylindrite, diadochite, diaphorite, franckeite, galena, gibbsite, jamesonite, kesterite, marcasite, metastibnite, miargyrite, natroalunite, orthoclase, plagionite, pyrite, quartz, rhodostannite, rhomboclase, semseyite, siderite, sphalerite, stannite, stibnite, tetrahedrite, tourmaline, valentinite, vivianite, wavellite, *wurtzite(?)*, zinkenite; see Itos and San José (the two most important mines), and Santo Cristo, Socavón, Antofagasta, Tetilla, San Felipe, La Colorada, Cerro San Pedro; page 26
Pacajake (aka the preferred Virgen de Surumi)	mine		2 km E of Hiaco Mine; 20 km NE of Colquechaca, Chayanta prov., Potosí	*Ahlfeldite*, barite, clausthalite, chalcomenite, cobaltomenite, grimaldiite, hematite, *mandarinoite*, molybdomenite, naumannite, *olsacherite*, *penroseite*, selenium, siderite, tiemannite; see Virgen de Surumi
Cañyon Pacopampa	mine		Caluyo, Inquisivi prov., La Paz	Ferberite, siderite
Pacuani	mountain		20 km E of Patacamaya, Aroma prov., La Paz	Bournonite, galena, jamesonite
Pacuni Chico	mine		Quime, Inquisivi prov., La Paz	Wolframite group minerals
Pailaviri	mine		Potosí, Bustillo prov., Potosí	part of the Cerro Rico complex; see Cerro Rico
Palca	river, village	16°34'S 67°59'W	on Illimani, 15 km ENE of La Paz, Murillo prov., La Paz	Stibnite; see Tajahahuira
Pallaya	mine		Yani, Larecaja prov., La Paz	Gold
Pampa Grande	district		35 km SE of Tupiza, Sud Chichas prov., Potosí	Aurichalcite, galena, linarite; see Churquina
Paragüi	mine		30 km NW of Colquiri, Inquisivi prov., La Paz	Chalcedony, hübnerite, stibnite
Patacamaya	town	17°14'S 67°55'W	80 km S of La Paz, Aroma prov., La Paz	see Pacuani, La Escosasa
Pazña	town, district	18°36'S 55°55'W	85 km SSE of Oruro, Poopó prov., Oruro	see Antequera, Avicaya, Totoral, Monserrat
Perez	mine		Huanuni, Dalence prov., Oruro	Wolframite group minerals

Left: *Tetrahedrite; 15 cm wide; Pulacayo (Huanchaca) Mine, Huanchaca, Quijaro province, Potosí; Canadian Museum of Nature collection; Michael Bainbridge photo*

Right: *Silver; 10 cm wide; Porco Mine, Quijarro province, Potosí; Terry Wallace collection; Jesse La Plante photo*

Place Name	Place Type	Lat. Long.	Province, Department	Known Mineral Species/ Notes
Pirpintuyok (aka Copacabana)	mine		Mochará, Sud Chichas prov., Potosí	Anglesite, galena
Poconota	mine		near Toropalca Mine, Nor Chichas prov., Potosí	Gold, stibnite
Polo Sur	mine		20 km NE of Sorata, Ancoma, Larecaja prov., La Paz	Cassiterite
Pongo	village		Murillo prov., La Paz	Cassiterite, ferberite, hydrotungstite, tungstite; see Martha, San Antonio 3, Alasca
Poopó	district, town	18°22'S 66°58'W	48 km SSE of Oruro, Poopó prov., Oruro	Cassiterite, *cylindrite*, fluorapatite, franckeite(?), "*incaite*", jamesonite, kermesite, silver, stannite, teallite, valentinite, vivianite, wurtzite, zinkenite; see Coriviri, San Francisco 3, Santa Cruz 2, Trinacria, Rosario 3, Bolivia, Candelaria; page 24
Porco	mine, town	19°50'S 65°59'W	35 km SW of Potosí, Quijarro prov., Potosí	Acanthite, ankerite, argyrodite, barite, calcite, cervantite, galena, marcasite, pyrargyrite, pyrite, pyrrhotite, quartz, siderite, silver, sphalerite, stephanite; see Tornohuico, Gutierrez
Porvenir 1	mine		Huanuni, Dalence prov., Oruro	Boulangerite, cassiterite ("needle tin"), cylindrite, franckeite, greenockite, freieslebenite, jamesonite, pavonite, plagionite, semseyite, sphalerite, stannite, stibiconite, teallite, valentinite, wurtzite
Porvenir 2 (aka Socavón Porvenir)	mine		adjacent to the Rosario 1 Mine in Asiento de Araca, Loayza prov., La Paz	Gold, tourmaline
Porvenir 3	mine		E. Sud Lípez prov., Potosí	incorrect name for Bolivar 1
Posokoni (Pozokoni)	mountain		near Huanuni, Dalence prov., Oruro	Cassiterite, jarosite, opal, scorodite, sphalerite, tourmaline; see Cataricagua
Potosí	city	19°35'S 65°45'W	Bustillo prov., Potosí	see Cerro Rico
Pucara	mine		Corocoro district, Pacajes prov., La Paz	Native copper
Pucro	mine		Chayanta, Bustillo prov., Potosí	Cassiterite, pyrite, pyrrhotite, siderite
Pueblo Viejo	mine		Esmoraca, Sud Chichas prov., Potosí	Bismutite
Puerto Suárez	town	18°56S 57°50'W	Busch prov., Santa Cruz	see La Gaiba, Mutún
Pulacayo (aka Huanchaca)	town, mine	20°23'S 66°42'W	Huanchaca, Quijaro prov., Potosí	Anhydrite, arsenolite, barite, bournonite, chalcostibite, enargite, freibergite (?), galena, geocronite, gypsum, jamesonite, miargyrite, metastibnite, pyrargyrite, pyrite, quartz, siderite, sphalerite, tennantite, tetrahedrite, wurtzite, zinkenite
Quechisla	town, subdistrict	20°54'S 66°1'W	64 km NNW of Tupiza, Nor Chichas prov., Potosí	Most minerals attributed to Quechisla actually come from the larger Animas-Chocaya district

Place Name	Place Type	Lat. Long.	Province, Department	Known Mineral Species/ Notes
Quetena Grande	town	22°18'S 67°23'W	97 km SW of San Pablo, Sud Lípez prov., Potosí	Metasideronatrite, malachite
Quiabaya	village	15°37'S 68°46'W	NE of Chichulaya, Larecaja prov., La Paz	Jamesonite; see Cascabel
Quime	town	16°59'S 67°13'W	10 km SSW of Inquisivi, Inquisivi prov., La Paz	Antimony, calcite, pyrite, scheelite valentinite, wolframite; see Chambillaya, Margareta, Pacuni Chico
Quimsacoya	mine		Huanuni dist., Dalence prov., Oruro	Hübnerite, stibnite
Quioma (aka Asiento)	mine		40 km SSW of Mizque, Mizqui prov., Cochabamba	Galena (Ag rich), litharge, minium, sphalerite
Ramiro	mine		San Pedro De Buena Vista, Charcas prov., Potosí	Cassiterite, pyromorphite
Recompensa	mine		San Ramón dist., Ñuflo de Chavez prov., Santa Cruz	Ferrocolumbite, monazite, muscovite
Remedios	mine		Corocoro, Pacajes prov., La Paz	Calcite
Rincón del Tigre	village		40 km NNW of Puerto Suarez, Busch prov., Santa Cruz	Chromium-rich chalcedony, quartz with 2-phase inclusions (bubbles)
Rosario 1 (aka Socavón Rosario)	mine		Asiento de Araca, Loayza prov., La Paz	Cassiterite, ferberite twins, gold, jamesonite, quartz with tourmaline inclusions, plagionite, scheelite; see Porvenir 2
Rosario 2	mine		western slopes of Cerro Bonete, Sud Lipez prov., Potosí	Pyrargyrite, silver
Rosario 3	mine	18°23'S 66°58'W	Poopó, Poopó prov., Oruro	Cylindrite, franckeite, sphalerite
Rosario 4 (aka Tasna-Rosario)	mine		Tasna, Sud Chichas prov., Potosí	Bismuth, bismuthinite, ferberite twins, quartz (Japan-law twins), pyrite
Rosita	mine		Cacachaca, Avaroa prov., Oruro	Chalcostibite
Sabaya	town, river, pampa	19°00'S 68°22'W	177 km SW of Oruro, Atahuallpa prov., Oruro	Ulexite, its probably near where the Sabaya River empties into the Salar de Coipasa
Sacaba	city	17°23'S 66°02'W	177 km SW of Oruro, Chapare prov., Cochabamba	Siderite
Salar de Empexa	salt flats		Campos prov., Potosí	see Empexa
Salar de Uyuni	salt flats		Potosí department	see Uyuni
Salo	village		20 km N of Tupiza, Sud Chichas prov., Potosí	see Tamacaya
Salvadora	mine		Oruro, Cercado prov., Oruro	see Socavón
Salvadora	mountain, stock		Bustillo prov., Potosí	This mountain hosts the Llallagua deposit. All minerals come from the Siglo XX Mine.
San Agustín 1	mine		3 km N of Corocoro Mine, Pacajes prov., La Paz	Copper after aragonite, cuprite, malachite
San Agustín 2	mine, village		35 km ESE of San Ramón, Ñuflo de Chavez prov., Santa Cruz	Columbite, magnetite, monazite, muscovite
San Antonio 1 (aka San Antonio de Calacalani)	mine		12 km S of Colquiri, Cercado prov., Oruro	Ferberite, *hydrotungstite*, stibnite, tungstite
San Antonio 2	mine		3 km from Viloco, Loayza prov., La Paz	Cassiterite; Some specimens from here may be labeled "Viloco."
San Antonio 3	mine		Pongo, Murillo prov., La Paz	Ferritungstite, tungstite, wolframite
San Antonio de Lípez	town	21°47'S 66°48'W	30 km SW of San Pablo, Sud Lípez prov., Potosí	Chlorargyrite; see Mesa de Plata, San Baldomero
San Baldomero	abandoned mine		Larecaja prov., La Paz	former name for Marcamarcani Mine
San Calixto	claim		Mt Huañaquino, Frías prov., Potosí	Apatite, feldspar, hematite, magnetite, rutile
San Carlos 1	mine		Chorolque, Sud Chichas prov., Potosí	Bismite
San Carlos 2	vein system		Chacaltaya Mine, Murillo prov., La Paz	Cassiterite, chalcopyrite, fluorite, gersdorffite, muscovite, pyrite, quartz, tourmaline
San Cristobal	town, mountain	21°07'S 67°12'W	90 km SSW of Uyuni, Nor Lípez prov., Potosí	Silver; see Cobrizos, Challahuilque
San Felipe	mine		Oruro, Cercado prov., Oruro	Augelite, andorite, pyrite, stannite
San Fermin	mine		S of San Ignacio and W of Santa Ana, Velasco prov., Santa Cruz	Quartz ("amethyst")
San Francisco 1	mine		Alto Chapare dist., Chapare prov., Cochabamba	Magnesioriebeckite and magnesite

Place Name	Place Type	Lat. Long.	Province, Department	Known Mineral Species/ Notes
San Francisco 2	mine		Corocoro district, Pacajes prov., La Paz	Copper minerals; see Corocoro
San Francisco 3	mine		Poopó dist., Poopó prov., Oruro	Cylindrite, kermesite
San Javier	town	16°20'S 62°38'W	60 km SW of Concepción, Ñuflo de Chávez prov., Santa Cruz	Tantalite
San José	mine		Oruro, Cercado prov., Oruro	Alunite, andorite, arsenopyrite, augelite, boulangerite, bournonite, cassiterite, cerussite, chalcopyrite, chlorargyrite, cylindrite, diaphorite, dickite, franckeite, galena, halotrichite, jamesonite, jarosite, kesterite, marcasite, metastibnite, plagionite, pyrite, rhomboclase, semseyite, stannite, stibnite, teallite, tetrahedrite, tourmaline, valentinite, vivianite, wurtzite, zinkenite
San José de Culini	mine		7 km E of Chalcaltaya, Sud Yungas prov., La Paz	Galena, pyrrhotite, siderite, sillimanite, ullmannite
San Luis 1 (aka Mary Mary, Heil Hitler, Kairiri)	mine	17°16'S 69°13'W	Berenguela dist., Pando prov., La Paz	Azurite, copper, galena, greenockite, sphalerite, tennantite
San Luis 2	mine		Chulchucani, Frías prov., Potosí	Asbolane, cuprian heterogenite
San Miguel 1	town		35 km S of San Ignacio de Velasco, Velasco prov., Santa Cruz	Staurolite
San Miguel 2	mine		Chorolque, Sud Chichas prov., Potosí	Wittichenite
San Miguel 3	mine	18°02'S 67°57'W	Lequepalca, Cercado prov., Oruro	Ferberite, jarosite, stibiconite, stibnite, sulfur
San Pablo de Churiña	mine		Macha, Chayanta prov., Potosí	see Churiña
San Pablo de Napa (aka Napa)	village (abandoned)	20°29'S 68°39'W	170 km WSW of Uyuni, Campos prov., Potosí	see Napa
San Pedro	village		10 km NW of Sorata, Larecaja prov., La Paz	see Gruta de San Pedro
San Pedro de Buena Vista	town	18°13'S 65°59'W	155 km N of Potosì, Charcas prov., Potosí	Barite, bismuth, bismuthinite, gold; see Santa María, Ramiro
San Ramón	town, district		80 km SW of Concepción, Ñuflo de Chavez prov., Santa Cruz	see La Bella, San Agustín 2, La Verde, El Paraiso, Recompensa
San Simón	hills		eastern Iténez prov., Beni	see Serranía San Simón
San Vicente	mine, town	20°15'S 65°33'W	40 km S of Atocha, Atocha-Quechisla dist., Sud Chichas prov., Potosí	Freibergite (?); Butch Cassidy and the Sundance Kid were allegedly killed here; see Alcacoya, Asunta, Soracaya, Tatasí
Santa Ana	town	16°37'S 60°43'W	35 km SE of San Ignacio de Velasco, Velasco prov., Santa Cruz	Beryl, limonite after almandite, muscovite, staurolite, tantalite
Santa Barbara	mine		Andacaba dist., Linares prov., Potosí	Wolframite group minerals
Santa Cruz 1	mine		between Caxata and Yaco, Loayza prov., La Paz	Mottramite
Santa Cruz 2	mine		Poopó, Poopó prov., Oruro	*Cylindrite*, franckeite, kermesite, zinkenite
Santa Fé	mine, district		3 km from Morococala, Dalence prov., Oruro	Cassiterite, crandallite, freieslebenite, matildite, miargyrite, muscovite, pyrargyrite, quartz, silver; see Negro Pabellon, Morococala
Santa Isabel	mine		Condeauque dist., Cercado prov., Oruro	Ferberite, hübnerite, zinkenite
Santa María	mine		San Pedro de Buena Vista, Charcas prov., Potosí	Bismuthinite, gold
Santa Marta	mine		12 km W of Aiquile, Campero prov., Cochabamba	Acanthite, galena, silver
Santa Rosa	vein		Antequera Mine, near Monserrat, Pazña dist., Poopó prov., Oruro	Siderite, teallite
Santiago	mine		Oploca 2, Sud Chichas prov., Potosí	Dickite
Santiaguillo	gully, village		15 km SE of Macha, Chayanta prov., Potosí	Adularia, cassiterite ("wood tin"), hematite, topaz
Santo Cristo	mine		Oruro, Cercado prov., Oruro	Andorite, augelite, aurichalcite, cassiterite, franckeite, jamesonite, metastibnite, pyrite, stibnite
Sayari	mountain		50 km SW of Cochabamba, Arque prov., Cochabamba	Cerussite, chlorargyrite, galena, manganese oxides

Place Name	Place Type	Lat. Long.	Province, Department	Known Mineral Species/ Notes
Serranía San Simón	hills	13°35'S 62°10'W	eastern Iténez prov., Beni	Hausmannite, "psilomelane", pyrolusite
Sica Sica	town	17°22'S 67°44'W	105 km SSE of La Paz, Aroma prov., La Paz	see Laurani, Lomitas, Lili
Siete Suyos	mine		Animas-Chocaya subdistrict, Sud Chichas prov., Potosí	Apatite, boulangerite, bournonite, freibergite, hocartite, jamesonite, pyrite, stannite
Siglo Veinte, Siglo XX	mine		Llallagua, Bustillo prov., Potosí	see Llallagua
Sijsihua	village	20°07'S 68°19'W	Empexa, Campos prov., Potosí	Nitratine
Socavón (aka La Salvadora 1)	mine		Oruro, Cercado prov., Oruro	Augelite, bournonite, franckeite, chalcostibite, plagionite, pyrite, sphalerite, stannite, stibnite, tetrahedrite, wavellite
Socavón Carmen	adit		Chojlla, Alonso de Ibanez prov., Potosí	Cassiterite, feldspar, muscovite, wolframites
Socavón Rosario	mine		Asiento de Araca, Loayza prov., La Paz	see Rosario 1
Soracaya	mine		10 km ESE of San Vicente, Atocha dist., Sud Chichas prov., Potosí	Pyrargyrite, tetrahedrite
Sorata	town, district	15°47'S 68°40'W	146 km NNW of La Paz, Larecaja prov., La Paz	see Millapaya, Hucumarini, Marcamarcani
Sucre	mine		10 km N of Esmoraca, Sud Chichas prov., Potosí	Gold, stibnite
Tacora	mountain		Avaroa, Nor Lípez prov., Potosí	Sulfur
Tajahahuira	mine		Palca, Murillo prov., La Paz	Gold and stibnite mine at 3,850 m elevation
Tamacaya	mine		10 km N of Salo, Sud Chichas prov., Potosí	Boulangerite
Tamiñani	elluvial deposit, mine		4 km N of Cerro Chicote Chico, Ayopaya prov., Cochabamba	Arsenopyrite, tungstite, wolframite group minerals
Tanapaca (aka Mocoya)	mine		3 km NW of Viloco, Loayza prov., La Paz	Arsenopyrite, bismuth, bismuthinite, called cassiterite, gold. Some specimens from here may be labeled "Viloco."
Tapacarí	town	17°31'S 66°36'W	45 km WSW of Cochabamba, Tapacarí prov., Cochabamba	Stibnite, sulfur; see La Sorpresa
Tapi	mine		Cotagaita, Nor Chichas prov., Potosí	Chalcostibite, stibnite
Tarabuco	town, region	19°10'S 64°52'W	64 km ESE of Sucre, Yamparáez prov., Chuquisaca	Amethyst, chalcedony, copper, cuprite, zeolites
Tarapaya	village, hot springs	19°28'S 65°47'W	15 km NNW of Potosí, Frías prov., Potosí	Hematite
Tarumita	mine		4 km NE of Caracollo, Cercado prov., Oruro	Celestine, cervantite, cinnabar on stibnite, tripuhyite
Tasna (aka Tazna)	mine, mountain	20°37'S 66°11'W	Atocha-Quechisla district, Sud Chichas prov., Potosí	Aheylite, apatite, arsenobismite, arsenopyrite, augelite, bindheimite, bismite, bismoclite, bismuth, bismuthinite, bismutite, cassiterite, *daubréeite*, ferberite twins, jamesonite, jarosite, joseite, preissingerite, pyrite, quartz (Japan-law twins), siderite, tungstite, vivianite, woodhouseite after apatite; see Constancia, Farellones, Mariana, Rosario 4; page 44
Tatasí	mine		25 km ENE of San Vicente, Sud Chichas prov., Potosí	Andorite, canfieldite (?), cerussite, miargyrite, ramdohrite, valentinite
Terremoto (aka Pepito)	mine		10 km N of Cacachaca, Avaroa prov., Oruro	Chalcostibite, stibnite
Tetilla	mine		Oruro, Cercado prov., Oruro	Anglesite, barite, galena, sphalerite, tetrahedrite
Tihua	mining company			operated Carguaicollo Mines
Tipuani	town, river	15°33'S 68°00'W	72 km NE of Sorata, Larecaja prov., La Paz	Cassiterite, garnet, gold nuggets, pyrite; see Chusi
Titina	mine		Caluyo, Inquisivi prov., La Paz	Tetrahedrite (2% Hg)
Tolapalca (aka Thola Palca)	mine		30 km ENE of Oruro, Cercado prov., Oruro	Ankerite, stibnite
Tomokoni	mine		Colavi dist., Saavedra prov., Potosí	Vivianite
Tornohuaico	mine		2 km S of Porco, Porco dist., Quijarro prov., Potosí	Acanthite, argyrodite/canfieldite, galena, silver, stannite, stephanite, siderite

Upper left: *Teallite; 2.1 cm tall; Santa Rosa vein, Antequera Mine, Pazña district, Poopó province, Oruro Arkenstone specimen (ex Philadelphia Academy) Joe Budd photo*

Middle left: *Silver; 3.5 cm tall; Veta (vein) Lipez, San Cristobal Mine, Nor Lipez province, Potosí George Witters collection; Jeff Scovil photo*

Lower left: *Bismuth nugget (polished); 5 cm wide Vilaque River, Murillo and Los Andes provinces, La Paz Mikon specimen; Jeff Scovil photo*

Upper right: *Pyrite; 6.2 cm tall; Ubina Mine, Quijarro province, Potosí; Tom Hughes collection; Jeff Scovil photo*

Left: *Valentinite; 4 cm tall; Tatasí Mine, Sud Chichas province, Potosí; Joan Massagué collection; Jesse La Plante photo*

Lower middle: *Vivianite; 2.5 cm tall; Huanuni Mine, Colavi district, Saavedra province, Potosí; Arkenstone specimen; Joe Budd photo*

Lower right: *Jamesonite on stannite; 2 cm wide; San José Mine, Oruro, Cercado province, Oruro; Tony Peterson collection and photo*

Place Name	Place Type	Lat. Long.	Province, Department	Known Mineral Species/ Notes
Toro Toro	town	18°07'S 65°46'W	27 km ENE of San Pedro, Charcas prov., Potosí	Calcite, descloizite
Toropalca	town	20°21'S 65°46'W	51 km N of Cotagaita, Nor Chichas prov., Potosí	Cassiterite, chalcostibite; see Poconota
Totoral	town, mine	18°50'S 67°00'W	Pazña district, Poopó prov., Oruro	Vivianite
Tras Cuarenta	mine		adjacent to the Viloco Mine, Loayza prov., La Paz	Ferrimolybdite, molybdenite, tourmaline and wolframite group minerals
Tricolore	mine		Don Diego, border of Frías and Saavedra provinces, Potosí	Chrysocolla
Trinacria	mine		6 km SSE of Poopó, Poopó prov., Oruro	Cylindrite, franckeite, zinkenite
Tupiza	city	21°24'S 65°54'W	201 km S of Potosí, Sud Chichas prov., Potosí	Almandine, epidote; see Salo
Turco	town	18°10'S 68°11'W	120 km WSW of Oruro, Sajama prov., Oruro	Azurite, chrysocolla, copper, cuprite; see La Ponderosa, Cuprita
Ubina	canton, mountain, mine	20°29'S 66°29'W	between Tasna and Pulacayo, Quijarro prov., Potosí	Barite, enargite, jamesonite, orthoclase, pyrite, tetrahedrite; see Moquegua, Orlandini, Cerro Ubina
Uncía	city	18°28'S 66°36'W	77 km SE of Oruro, Bustillo prov., Potosí	All specimens ascribed to Uncía except pyrargyrite, psilomelane and common opal, are from the Siglo XX Mine; see Amayapampa, Salvadora
Unificada	mine		Potosí, Bustillo prov., Potosí	see Cerro Rico
Urania	mine		S slope of Illimani, Murillo prov., La Paz	Ferberite after scheelite, wolframite
Uturuncu	volcano and hot springs area	22°13'S 67°14'W	Sud Lípez prov., Potosí	Alunogen, opal, pickeringite, pyrolusite, pink sulfur
Uyuni	salt flats		Potosi department	halite, ulexite
Valle Zongo	valley		Larecaja and Murillo provinces, La Paz	see Zongo valley
Venus	vein system		near, or part of the Antequera Mine, Poopó prov., Oruro	Cassiterite, hydrocerussite, varlamoffite
Vetillas	mine		Maragua, Chayanta prov., Potosí	Stannite, franckeite
Víboras	mine		Machacamarca, Saavedra prov., Potosí	Bournonite, chlorargyrite, tetrahedrite
Vichacla	village	20°38'S 65°38'W	Nor Chichas prov., Potosí	see Cerro Negro 2
Vicuñita	prospect		4 km S of the María Teresa Mine, Avaroa prov., Oruro	Cassiterite ("needle tin"), galena, teallite
Vila Apacheta	mine	19°55'S 66°16'W	near Chayanta, Bustillo prov., Potosí	Apatite, cassiterite, gypsum, *rhodostannite*, magnesite, pyrite, sphalerite, stannoidite
Vilaque	river		Murillo and Los Andes provinces, La Paz	Alluvial bismuth, bismuthinite, gold, quartz
Villa Tunari	town		90 km NE of Cochabamba, Chapare prov., Cochabamba	see Alto Chapare
Viloco (aka Araca)	mine	16°53'S 67°37'W	41 km NNE of Luribay, Loayza prov., La Paz	Apatite, arsenopyrite, cassiterite, chamosite, molybdenite, ferrimolybdite, pyrite, pyrrhotite, quartz, siderite, sphalerite, tin, tourmaline, wolframite; see San Antonio 2, Tanapaca
Virgen de Surumi	mine		Pacajake, Chayanta prov., Potosí	preferred name for Pacajake Mine
Volcano	mine		Mojo, Omiste prov., Potosí	Cerussite, pyromorphite, wulfenite
Yaco	town	17°09'S 67°24'W	29 km ESE of Luribay, Loayzo prov., La Paz	see Santa Cruz 1
Yanacachi	town, subdistrict	16°23'S 67°43'W	Mururata dist., Sud Yungas prov., La Paz	see Chojlla, Bolsa Negra, El Barco
Yani	town	15°36'S 68°35'W	20 km NNE of Sorata, Larecaja prov., La Paz	see Pallaya
Yarvicoya	mine	17°37'S 66°55'W	40 km NE of Oruro, near Lequepalca, Cercado prov., Oruro	Azurite, brochantite, chalcanthite, chalcopyrite, copper, cuprite, malachite, stibiconite, stibnite
Yura	town	20°04'S 66°08'W	40 km SSW of Porco, Quijarro prov., Potosí	see Caracota
Zongo (includes the Fabulosa Mine)	valley		45 km N of La Paz, Larecaja and Murillo provinces, La Paz	Almandine, apatite, biotite, cassiterite, chalcopyrite, chlorite, erythrite, hübnerite, lazulite, molybdenite, muscovite, quartz, sphalerite, stannite, triphyllite, triplite, vivianite; home to at least 30 mines

KRISTALLE est. 1971
Wayne & Dona Leicht • 875 North Pacific Coast Highway, Laguna Beach, CA 92651-1415, USA
phone (949) 494-7695 & -5155• fax (949) 494-0402 • email leicht@kristalle.com • www.kristalle.com
We buy collections! Over 150 collections purchased since 1971!
Photo by Harold & Erica Van Pelt
CRYSTAL CLASSICS
Ian Bruce • PO Box 3967, Yeovil, Somerset, England BA20 9AH
phone (+44) 1935-479822 • email ian@crystalclassics.co.uk • www.crystalclassics.co.uk

References

Abbott, M. B. & Wolfe A. P. 2003: Intensive Pre-Incan Metallurgy Recorded by Lake Sediments from the Bolivian Andes. *Science* **301**/5641: 1893-1895.

Ahlfeld, F. 1936: The Bolivian tin belt. *Economic Geology* **31**. 48-72.

Ahlfeld, F. 1953: Die Metallprovincen Altiplano (Bolivien). *Neues Jahrbuch Mineralogie Abhandlungen* **85**: 1-58.

Ahlfeld, F. 1954: *Los yacimientos minerales de Bolivia.* Bilbao, Imprenta Industrial, S. A., 227 p.

Ahlfeld, F. & Muñoz Reyes, J. 1943: *Los minerales de Bolivia.* Direction General de Minas y Petroleo, La Paz, 276 p.

Ahlfeld, F. & Muñoz Reyes, J. 1955: *Las especies minerales de Bolivia.* Banco Minero de Bolivia, La Paz, Bolivia. 180 p.

Ahlfeld, F. & Schneider-Scherbina, A. 1964: *Los yacimientos minerales y de hidrocarburos de Bolivia.* Departamento National de Geologia, Bolivia, Boletin **5**, 388 p.

Ahlfeld, F. & Mosebach, R. 1935: Über Alkaligesteine in der bolivianische Ostkordillere. *Neues Jahrbuch für Mineralogie* **69A**: 388-414.

Ahlfeld, F. & Wegener, R. 1931: Über die Herkunft der im Bereich altperuanischer Kulturen gefundenen Schmuckstücke aus Sodalith. *Zeitschrift für Ethnologie* **63**: 288-296.

Amstutz, G. C. 1981: Kupfer nach Aragonit. *Lapis* **6**/11: 19.

Arce Burgoa, O. R. 2007: *Guia a los Yacimientos Metaliferos de Bolivia.* Minera San Cristobal, La Paz. 307 p.

Atkinson, D. 2005: Following the Trail of Butch Cassidy and the Sundance Kid in Bolivia. http://www.gonomad.com/features/0507/butch_cassidy_trail_in_bolivia.html

Aurelia, M. 1994: Bolivia http://ddg.com/LIS/aurelia/bolivi.htm

Back, M. E. & Mandarino, J. A. 2008: *Fleischer's Glossary of Mineral Species.* The Mineralogical Record. 345 p.

Bandy, M. C. 1944: *Mineralogy of Llallagua, Bolivia.* La Paz. 1976 reprint by Tucson Gem and Mineral Society as *Special Publication No. 1.*

Barba, A. A. 1640: *El Arte de los Metales.* Madrid (reprint Potosí 1967).

Berton, A. 1936: The Corocoro Copper District of Bolivia. *American Institute of Mining, Metallurgical and Petroleum Engineers, Techn Publication* **698**, 1-17.

Berton, A. 1937: The Corocoro Copper District of Bolivia. *Transactions of the Society of Mining, Metallurgical and Petroleum Engineers, Inc. (AIME)* **126**: 541-558.

Brendler, W. 1932: Über Sodalith vom Cerro Sapo. *Zentralblatt für Mineralogie Abteilung A, Mineralogie und Petrographie* 42-46.

Brendler, W. 1934: Sodalith from Bolivia. *American Mineralogist* **19**: 28-31.

Campbell, D. F. 1942: The Oruro silver-tin district, Bolivia. *Economic Geology* **37**: 87-115.

Chace, F. M. 1948: Tin-silver veins of Oruro, Bolivia. *Economic Geology* **43**: 333-383, 435-470.

Clark, A. H. 1972: On the natural occurrence of tin sulfides (berndtite). *Naturwissenschaften* **59**/8: 361

Cook, R. B. 1975: The Mineralogy of the Department of Oruro, Bolivia. *The Mineralogical Record* **6**/3: 125-137.

Cook, R. B., 1977, Morphology and Occurrence of Bolivian Cassiterite. *The Mineralogical Record* **8**/1: 52 – 57

Domeyko, J. 1880: Cristeaux epigenes de cuivre metallique de Corocoro. *Analles de Mines* **7**: 531-537.

Entwistle, L. P., & Gouin, L. O. 1955: The Chalcocite-ore Deposit at Corocoro, Bolivia. *Economic Geology* **50**: 555-570.

Evans, D. L. 1940: Structure and mineral zoning of the Pailaviri section, Potosí, Bolivia. *Economic Geology* **35**: 737-750.

Foord, E. E. & Taggart, J. E. 1998: A reexamination of the turquoise group: the mineral aheylite, planerite (redefined), turquoise and coeruleolactite. *Mineralogical Magazine* **62**: 93-111.

Gordon, S. G. 1923: Vauxite and Paravauxite, two new minerals from Llalagua, Bolivia. *Proceedings of the Academy of Natural Sciences of Philadelphia* **75**: 261-267.

Gordon, S. G., 1924: Crystallographic notes on six minerals from Peru and Bolivia. *Proceedings of the Academy of Natural Sciences of Philadelphia* **76**: 335-341.

Gordon, S. G. 1939: Thorium-free monazite from Llallagua, Bolivia. *Academy of Natural Sciences of Philadelphia, Notulae Naturae*, no. 2 (7pp).

Gordon, S. G. 1944: The Mineralogy of the Tin Mines of Cerro de Llallagua, Bolivia. *Proceedings of the Academy of Natural Sciences of Philadelphia* **96**: 279-359.

Grice, J. D., Ercit, T. S., Hawthorne, F. C. 1993: Povondraite, a redefinition of the tourmaline ferridravite. *American Mineralogist* **78**: 433 - 436

Groesbeck, G. A. P. 2007: A Brief History of the Jesuit Missions of Chiquitos (Eastern Bolivia). *Bolivian Studies Journal* **14**: 99-120.

Grundmann, G., Lehrberger, G., Schnorrer-Kohler, G. 1990: The El-Dragon Mine, Potosí, Bolivia. *The Mineralogical Record* **21**/2: 133-146.

Guillermoprieto, A. 2008: Bolivia's New Order. *National Geographic* **214**/1: 88-103.

Heilmeir, H., 1930: Über Zinnstein in Formen and Akzessorien. *Neues Jahrbuch für Mineralogie Abhandlungen* **61**: 403-468.

Herzenberg, R. 1933: Gumucionit, eine neue arsenhaltige Varietät der Schalenblende. *Zentralblatt für Mineralogie, Geologie und Paläontologie, Abteilung A, Mineralogie und Petrographie* **2**: 77-78.

Huminicki, D. M. C., Hawthorne, F. C., Grice, J. D., Roberts, A. C. & Jambor, J. L. 2003: Nikischerite—A New Mineral from the Huanuni Tin Mine, Dalence Province, Oruro Department, Bolivia. *The Mineralogical Record* **34**/2: 155-158.

Hurlbut, C. S. & Honea, A. R. 1962: Sigloite, a new mineral from Llallagua, Bolivia. *American Mineralogist* **47**: 1-8.

Hyršl, J. 2002: Cassiterite from Viloco, Bolivia. *Gems & Gemology* **38**/2: 175-176.

Hyršl, J. & Petrov, A. 1996: Potosí und Oruro—die bedeutendsten Silbererz-Lagerstätten Boliviens. *Mineralienwelt*
Hyršl, J. & Petrov, A. 1997: Eine der grössten Kupferlagerstätten der Welt: Corocoro in Bolivien. *Mineralienwelt* **8**/6: 30-35.

Weinrich

Dealer in Fine Mineral Specimens

Collections Bought Complete or Partial

Visit our weekly Auctions & Galleries **danweinrich.com**

Open every day of the week on the Internet

New FRIENDLY Search Engine

East Coast Gem, Mineral & Fossil Show, West Springfield, MA August 7–9, 2009

Denver Gem and Mineral Show, Denver, CO Booth J37 September 18–20, 2009

Holiday Inn, Denver, CO Room 109 September 13–20, 2009

Silver 8.3 cm, Peru — Jeff Scovil photo

Dan & Diana Weinrich
P.O. Box 425
Grover, Missouri 63040
314-341-1811
www.danweinrich.com

Hyršl, J. & Petrov, A. 1998 (a): Pseudomorphs from Bolivia, a review. *Rocks & Minerals* **73**/6: 410-414.

Hyršl, J. & Petrov, A. 1998 (b): Gemstones and ornamental stones from Bolivia: a review. *Journal of Gemmology* **26**/1: 41-47.

Hyršl, J. & Petrov, A. 1999: Sodalite from Bolivia. *Canadian Gemmologist* **20**/2: 54-56.

Hyršl, J. & Petrov, A. 2006: Famous Mineral Localities: Llallagua, Bolivia. *The Mineralogical Record* **37**, 2, 117-162.

Jaskolski, S. 1933: Les gisements argento-stanniferes de Potosí en Bolivie. *Archiwum Mineralogiczne* **9**, 47-92.

Kampf, A. R. 1982: Jeanbandyite, a new member of the stottite group from Llallagua, Bolivia. *The Mineralogical Record* **13**/4: 235-239.

Keutsch, F. & de Brodtkorb, M. K. 2008: Metalliferous parageneses of the San José mine. *Journal of South American Earth Sciences* **25**/4: 485-491

Language Crossing: History of Bolivia http://www.languagecrossing.com/Destinations/Bolivia/History_of_Bolivia

Laurs, B. M. (ed.) 2001: Update on amethyst, citrine and ametrine from the Anahí mine, Bolivia. *Gems & Gemology* **37**/4: 334-335.

Lindgren, W. & Abbott, A. C. 1931: The silver-tin deposit of Oruro, Bolivia. *Economic Geology* **26**: 433-479.

Lincoln, F. C. 1917: Oruro tin-silver district, Bolivia. *Min Sci Press* **115**: 57-58.

Mayell, H. 2003: Bolivia Silver Mines May Predate Inca, Experts Say, *National Geographic News*-Sept. 25, 2003, http://news.nationalgeographic.com/news/2003/09/0925_030925_silvermines.html

McCarthy, J. 2007: Bolivian Miners Split on Nationalization Plan. *National Public Radio*-March 31, 2009. http://www.npr.org/templates/story.php?storyID=7181006

Meadows, A. & Buck, D. 2001: Butch Cassidy & The Sundance Kid. http://ourworld.compuserve.com/homepages/danne/slatta.htm

Microsoft Corporation. Bolivia. Microsoft Encarta® online Encyclopedia 2008: http://encarta.msn.com/encyclopedia_761563800/Bolivia.html.

Miller, B. L. & J. T. Singewald 1919: *Mineral Deposits of South America.* 598 p. McGraw-Hill, New York.

Moëlo, Y.; Makovicky, E. & Karup-Moller, S. 1989: Sulfures complexes plombo-argentiferes: Mineralogie et cristallochimie de la serie andorite-fizelyite. *Document du BRGM* **167**, 107 p.

Moh, G. H. 1966: Das Binäre System Zinn-Schwefel und seine Minerale. *Fortschritte der Mineralogie* **42**: 211.

Moh, G. H. & Berndt F. 1964: Two new natural tin sulphides, Sn_2S_3 and SnS_2. *Neues Jahrbuch Mineralogie, Monatshefte* **3**: 94-95.

Moore, T. P. 2008: A brief history of early silver mining in Spanish America. *The Mineralogical Record* **30**/6: 6, 5-21.

Neumann-Redlin, C. 2007: Friedrich Ahlfeld. *Monatsblatt Centro Cultural Aleman* (La Paz) **1**.

Peele, R. 1894: The Silver Mines of Colquechaca, Bolivia. *Engineering and Mining Journal* **57**: 78-79, 100-101.

Petrov, A. 1994: Der neue Turmalin aus dem tropischen Regenwald Boliviens. *extraLapis* **6**: 50-53.

Petrov, A. 1996: Über Mineral-Vorkommen in Bolivien—Eine Übersicht. *Mineralienwelt*, **7**/6: 25-35.

Petrov, A. 2003: The El Desierto sulfur mine, Potosí, Bolivia. *The Mineralogical Record* **34**/4: 297-305.

Petrov, A. & Hyršl, J. 2008: Mineralogy of Alto Chapare, Cochabamba Department, Bolivia. *Rocks & Minerals* **83**: 114-123.

Petrov, A. & Smith, B. 2001: A guide to mineral localities in Bolivia. *The Mineralogical Record* **32**/6: 457-482.

Petrov, A. & White, S. 2001: Kylindrit von der Itos Mine, Oruro, Bolivien. *Lapis* **26**/9: 27-29.

Pohs, H. A. 1995: *The Miner's Flame Light Book*. Denver, CO Flame Publishing Company, 867 p.

Prior, G. T. & Spencer, L. J. 1897: The identity of andorite, sundtite, and webnerite. *Mineralogical Magazine* **11**: 286 – 301.

Rakovan, J. 2003: Exceptional Apatites from the Siglo XX Mine, Llallagua, Bolivia. *The Mineralogical Record* **34**/1: 117.

Read, J. 2008. *The Rough Guide to Bolivia*. Rough Guides, New York, London, Delphi. 389-408.

Redwood, S. D. 2003: "The Father of Bolivian Geology" Friedrich (Federico) Ahlfeld (1892-1982). *The Mineralogical Record* **34**/3: 225-233.

Redwood, S. D. 2003: Famous Mineral Localities: Pacajake Selenium Mine, Potosí, Bolivia. *The Mineralogical Record* **34**/4: 339-357.

Rivas, S. & Ahlfeld, F. 1998: *Los Minerales de Bolivia y sus Parajes.* Talleres de Industrias Gráficas, Santa Cruz, Bolivia. 2 vols, 629 p.

Schultz, F.; Lehmann, B.; Tawackoli, S.; Rössling, R.; Belyatsky, B. & Dulski, P. 2004: Carbonatite diversity in the Central Andes: the Ayopaya alkaline province, Bolivia. *Contributions to Mineralogy and Petrology* **148**: 391-408.

Sillitoe, R. H., Halls, C. & Grant, J. N. 1975: Porphyry tin deposits in Bolivia. *Economic Geology* **70**: 913–927

Turneaure, F. S. 1935: The tin deposits of Llallagua, Bolivia. *Economic Geology* **30**: 14-70.

Turneaure, F. S. 1960: A comparative study of major ore deposits of central Bolivia. *Economic Geology* **55**: 217-254, 574-606.

Turneaure, F. S. 1971: The Bolivian Tin-Silver Province. *Economic Geology* **66**: 215-225.

U.S. Geological Survey & Servicio Geologico de Bolivia1992: *Geology and Mineral Resources of the Altiplano and Cordillera Occidental, Bolivia.* USGS Bulletin 1975. 365 p.

Vasconcelos, P. M., Wenk, H. R., Rossman, G. R. 1994: The Anahi Ametrine Mine, Bolivia. *Gems & Gemology* **30**/1: 4-23.

Virreira, W. H. 1979: *Breve historia de la mineria en Bolivia.* La Paz.

Walenta, K. & Dunn, P. J. 1977: Ferridravite, a new mineral of the tourmaline group from Bolivia. *American Mineralogist* **64**: 945-948.

Wallace, T. C. & M. K. Hall M. K. 2003: Minerals of the Andes: emeralds, gold and silver from the sky. *Rocks & Minerals* **78**/1: 12-38.

Wallace, T. C., Barton, M., Wilson, W. E. 1994: Silver & Silver-Bearing Minerals. *Rocks & Minerals* **69**/1: 16-38.

Weldon, R. 2009: Anahí's "new" ametrine. *Gems & Gemology* **45**/1: 63-64.

Whittaker, E. J. W. 1949: The structure of Bolivian crocidolite. *Acta Crystallographica* **2**: 312 ff.

Wilson, W. E. & Petrov, A. 1999: Famous Mineral Localities: Cerro Rico de Potosí, Bolivia. *The Mineralogical Record* **30**/1: 9-36.

Zacek, V., Petrov, A. & Hyršl, J. 1998: Chemistry and origin of povondraite-bearing rocks from Alto Chapare, Cochabamba, Bolivia. *Journal of the Czech Geologic Society* **43**/1-2: 59-67.

Zacek, V., Fryda, J., Petrov, A. & Hyršl, J. 2000: Tourmalines of the povondraite -(oxy)dravite series from the cap rock of meta-evaporite in Alto Chapare, Cochambamba, Bolivia. *Journal of the Czech Geologic Society* **45**/1-2: 3-12.

This incredible Green Fluorite cluster from Riemvasmaak, South Africa embodies aesthetic perfection. Such intensely saturated color on a matrix paved with amethyst crystals makes it a wonder to behold. The unusually fine luster makes it a star among all the Riemvasmaak fluorites.

Another example of the quality collectors from around the world have grown to expect from us.

FINE MINERALS

INTERNATIONAL

www.FineMineral.com

Scovil Photography

Vivianite with ludlamite on pyrite; 10.7 cm high; Huanuni Mine (160 meter level), Dalence, Oruro, Bolivia; Steve & Clara Smale collection

Scovil Photography · P.O. Box 7773 · Phoenix, AZ 85011
Phone (602) 254-0735 · Fax (602) 307-0394
email: jeffscovil@earthlink.net

Mineralientage
München
Oct. 30 - Nov. 1, 2009
INDIA
Crystals & Gems
DMF
INTERNATIONAL
recommended
HALL A6/A5/B6/B5
46th int. Show for
Minerals, Fossils, Gems
Jewelry and Accessories
Grafik PIELSTICKER Foto: J. Scovil
Europe's Top Show
GEOFA Dealers Day Fr. Oct. 30 Pre-registration requested. BÖRSE Public Fair Sa. Oct. 31 / Su. Nov. 1
www.mineralientage.com

WULFENITE ON MIMETITE; SAN FRANCISCO MINE; 6 CM WIDE; TONY POTUCEK COLLECTION; JEFF SCOVIL PHOTO

1441 WEST 46TH AVENUE; UNIT 8, DENVER, COLORADO 80211, USA
1-303-477-6897; DAVE@DAVEBUNKMINERALS.COM

DAVEBUNKMINERALS.COM